Indian Pest Aphids

Indian Pest Aphids

– *by* –

Dr. T.V. Sathe
Entomology Division
Department of Zoology
Shivaji University, Kolhapur, M.S.

&

Dr. B.V. Jadhav
Head, Department of Zoology
Bahasaheb Desai College, Patan

2018
Daya Publishing House®
A Division of
Astral International Pvt. Ltd.
New Delhi - 110 002

First Published, 2008
Reprinted, 2018

ISBN: 978-81-7035-544-1 (Hardbound)

Publisher's note:

Every possible effort has been made to ensure that the information contained in this book is accurate at the time of going to press, and the publisher and author cannot accept responsibility for any errors or omissions, however caused. No responsibility for loss or damage occasioned to any person acting, or refraining from action, as a result of the material in this publication can be accepted by the editor, the publisher or the author. The Publisher is not associated with any product or vendor mentioned in the book. The contents of this work are intended to further general scientific research, understanding and discussion only. Readers should consult with a specialist where appropriate.

Every effort has been made to trace the owners of copyright material used in this book, if any. The author and the publisher will be grateful for any omission brought to their notice for acknowledgement in the future editions of the book.

Published by : **Daya Publishing House®**
A Division of
Astral International Pvt. Ltd.
– ISO 9001:2008 Certified Company –
4736/23, Ansari Road, Darya Ganj
New Delhi-110 002
Ph. 011-43549197, 23278134
E-mail: info@astralint.com
Website: www.astralint.com

Preface

Aphids are very destructive insects of agricultural, horticultural and forest crops. They belong to order–Hemiptera of family aphididae. About 4000 species of aphids have been described from all over the world. It is believed that aphids are evolved 280 million years ago in the carboniferous period. They are soft bodied, louse like, winged or non winged insects and always found with large populations. Aphids suck the cell sap from plants and inject toxins, as a result leaves become curly, they turn yellow, and drop down. Similarly, flowering and fruiting bodies of plants drop down. They secrete honeydew like substance on leaves which create sooty moulds over leaf surface and affect photosynthesis, growth and finally the yield of crop. They also act as vector for viral diseases of crops. In Maharashtra, the sugarcane wooly aphid has paralized the entire sugarcane industry by destroying the sugarcane. Thus, aphids are very economically important insects. Therefore, any advance knowledge on their distribution, taxonomy, life cycle, host records, nature of damage and ecofriendly control measures will add great relevance in their pest management strategies. In present text, attempts have been made on 20 new species and 30 existing pest aphid species with respect to above mentioned aspects.

The information on aphid characterization and control measures is extremely useful for protecting the crops. We hope that the book will be useful guide to students, teachers, scientists, farmers, agriculturist and silviculturist both in India and abroad.

T.V. Sathe

B.V. Jadhav

Contents

Preface

1. General Introduction 1
2. Biology of Aphids 4
3. Crop Aphids and their Control 10
4. New Aphids on Crops and their Control 33

Bibliography 187

Index 205

Chapter 1

General Introduction

Indian agriculture is backbone of Indian economy. According to Singh and Sadhu (1986), India cultivates a wide varicty of food and non-food crops. Food crops refer to cereals, pulses, oil seeds, vegetables, fruits and other crops. In last fifty years, cropping intensity increased from 1.11 to 1.36 (Paroda, 1993). There are several significant achievements in different crops relating to breeding either genetically higher in productivity or possessing resistance to various biotic and abiotic iraces. India is the first in the world to commercialise hybrid cotton, H-4 was the first such interspecific hybrid. She is also first in the world to release commercial hybrids in castor and jowar. The compound growth rate of various crops during the last fifty years has been fairly increased. However, the expected yield of the crops is not achieved so far because of the insect pest damage. High yielding new hybrid varieties in several agricultural crops are also responsible for increasing the pest complex in agroecosystems. A very large number of pests inflict the damage to the stems, leaves, flowers, seeds and roots of agricultural plants. Many soil inhabiting insects damage roots and kill the plants causing serious losses in the agricultural and forest crops. Several insects cause damage to agricultural and forest crop by boring, the roots, stems and fruits. Likely, many insects acts as defoliators (which feed upon the leaves).

Leaf mining feeding of crop insects is also largely seen in agro and forest ecosystems. Gall insects are also not uncommon in above ecosystems. Likely, cell sap sucking insects also cause severe damage to the crops in above ecosystems. The insect pests like aphids, jassids, delphacids, psyllids, membracids cause enormous damage to agricultural and forest crops.

The suborder Homoptera contains some of the more important families like Aphididae, Coccidae, Pseudococcidae, Jassidae, Psyllidae, etc. which include large number of pest species and cause the damage to various plants by sucking the cell sap.

Aphids belongs to the family Aphididae of sub-order Homoptera of order Hemiptera and characterised by their minute (1-10mm) soft body and wingless or winged forms, abdomen with a pairs of cornicles on 6th abdominal segment at the posteriolateral region. They are often found in numerous colonies clustering over the tender parts of various plants. Several or all generations comprise parthenegenitic females which do not require fertilization and are viviparous. Winged aphids are known as *alatae* and wingless aphids, as *apterae*. Polymorphism is seen within species of different forms or morphs, is also characteristic of aphids.

The aphid antennae have two short thick basal segments and thinner flagellum, of at most four segments. There is an ocular tubercle made up of three lenses. (triommatidium) situated behind each compound eye. The wings have only one prominent longitudinal vein. There are two tarsal segments. Aphids, adelgids and phyllozerids belong to insect superfamily Aphidoidea. It is believed that Aphidoidea evolved 280 million years ago in the carboniferous period in paleozoic era. Aphidinae, largest subfamily of modern aphids is not represented in the fossil record until late tertiary and is associated with evolution of angiosperms, in particular the rosaceae.

While sucking the cell sap the aphids secrete honeydew like substances, later it is responsible for creating sooty moulds on the leaves. The sooty mould interferes with the photosynthetic activities of the crop. In many cases, as result of sucking the cell sap, leaves become curly, turn pale yellow, dry up and flowers and fruits drop down at pre-maturity. The insects are also responsible for causing nacrosis and transporting pathogenic diseases towards the crops. The important species recorded on different crops refers to *Rhophalosiphum maidis* (Fitch) on Sorghum (*Sorghum vulgare*, Pers) and Maize (*Zea mays* Linn.); *Aphis craccivora* (Koch) on cow pea (*Vigna catjanag* Walp) and on groundnut (*Arachis hypogea* Linn.); *Dactynotus compositae* (Theobald) on safflower (*Carthamus tinctorius* Linn.) and on Soyabeen (*Glycine soja* Linn.) and guava (*Psidium guavjava* Linn.). (Sathe 1992, 2004; Annonymous 1972).

There are about 4000 described species of aphids. Greatest number of aphid species occurs in temperate region and there one plant species in four is infested. Aphids are predominant in temperate region but recorded from all over the world. Within the Aphidinae and Drepanosiphine, that make up 70 per cent of modern Aphids, are not restricted to a particular region, but the Greenideinae and Hormaphidinae, that make up 7 per cent restricted mostly to South East Asia and Australia.

Small size is a feature of aphid ever since their evolution in the permian. Even, though there is a range of size within and between species. In general, bigger organisms have a better chance of reaching a reproductive condition and tend to produce more gametes (Calow, 1978). Therefore, it is of interest to consider that what determines size in aphids? as part of avenue for future research.

The review of literature indicates that aphids have been studied by Agarwala (1980), Basu (1961 a& b, 1964, 1967), Behura (1963), Cottier (1953), Das (1918), Eastop (1958, 59, 77), Ghosh et al (1970, 71), Raychaudhuri (1956, 1973, 1980), Smith (1972), Takahashi (1921, 24, 25), Theobald (1915), etc. from different parts of the world. However, Maharashtra have been totally neglected, except the work of Sathe (1992) and Jadhav and Sathe (2006a, b). Sathe (1992) reported 28 species of aphids on economic plants from Western Ghats. Similarly, Jadhav and Sathe (2006a, b) recorded 21 species of aphids from Western Ghats of Satara district and 19 species from Western Ghats of Poona district.

Aphids are thus economically very important group of insects since they cause the damage to various economic plants by sucking the cell sap and forming the galls and transmitting many viral diseases. Therefore, detail information on their distribution, characters (taxonomy), life cycle, reproduction, nature of damage, host plants and control measures is must. The present book will add great relevance on this contest.

Chapter 2

Biology of Aphids

Biological pecularities of an organisms help in designing proper control measure for the pest species. Therefore, knowledge on biology of an aphid species will add great relevance in pest management strategies. An aphid life cycle is made up of a sequence of morphs that differ in both their behaviour and structure. In a few species these can be as many as eight morphs that differs at least in their external morphology. Polymorphism is characteristics of aphids and reaches it greatest development in host alternating species.

Individual aphids are capable of defending themselves against a wide range of natural enemies with varying degrees of success. Depending on the relative size of the attacking parasites or predator, an aphid will kick them and attempt to walk away. A drop of the plant or smear of the wax can be exuded from its siphunculi over an attacker in certain cases (Dixon, 1958).

Aphids are known for their tremendous reproductive potential. According to Huxley (1858), if all the offspring survive than in the course of a season the descendants of one female contain more substance than 5,00,000,000 stout men. Some aphids produce hundreds of offspring. However, certain morphs, for example oviparae of the Pemphiginiae, produce only one. Early generation of apterae in particular those that hatch from the over wintering eggs, tend to be larger and more fecund than their descendants. The extreme examples are the fundatrices of some gall forming species that are nothing more than awkward bags' packed with embryos (Mordwilko, 1935, Shaposhnikov, 1977).

Aphids can propagate without fertilization and continue to do so as many as 10 generations (Bonnet, 1745). Under certain conditions, winged or wingless males appear and copulate with wingless oviparous female. This is the cyclical parthenogenesis characteristic of most aphids. The sexual reproduction is also the characteristic of aphids.

Aphid life cycle is of two types, autoecious (host-specific) and heteroecious (host alternating). Autoecious species of aphid live on one or a few species of a particular genus of plants while, heteroecious aphids spend autumn, winter and spring on a primary woody host and the summer usually on secondary herbaceous plants. Aphids are called as polyphagous when they feed on primary and secondary hosts belong to different families of plants. However, many aphids are sequentially monophagous as they live on one plant species at a time, for example, the damson hop aphid alternates between damson and hop.

According to Eastop (1977) members of four sub-families Aphidinae, Anoeciinae, Hormaphidinae, and Pemphiginae of the family Aphididae show host alternation (Eastop, 1977). Life cycle (Figures 1 a& b) of aphids is associated with some of/all of the following phenomena -

(*i*) Parthenogenesis, (*ii*) Vivipary (ovovivipary where in the egg hatch out while inside the uterus), (*iii*) scarcity or lack of males, (*iv*) polymorphism with winged and wingless adults (mouth parts/ sense organ well developed or poorly developed in individuals) and (*v*) alternation of host plants, based primarily on the last phenomenon.

Aphids are divided into three categories *viz.* (*i*) those that are widely polyphagous but devoid of migration (*e.g.* alternation of host), (*ii*) those that have restricted polyphagy but have facultative migration and (*iii*) those that are only slightly polyphagous but have obligatory migration. Various stages in the life history of an aphid are shown in Figures 1 a& b.

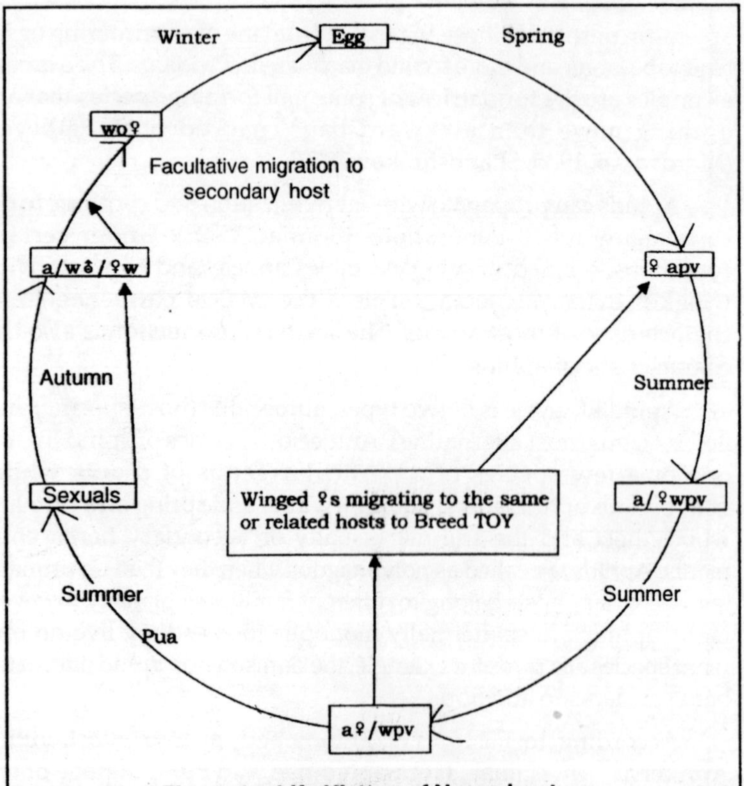

Figure 1a: Life History of Non-migratory and Facultative Aphids

In non-migratory aphids, the wingless, parthenogenetic and viviparous females continue to breed parthenogentically throughout the year. After producing several similar generation, on the some host, some of the individuals are born winged and they fly to different host plants of the some species. In case of aphids which have facultative migration, after several wingless, parthogenetic and viviparous generations, sexuals are produced by the end of summer which produce sexually reproducing males (winged and wingless) and female (winged) which mate in autumn to produce winged but oviparous females which lay eggs in winter on the same host plant or plants of the same species.

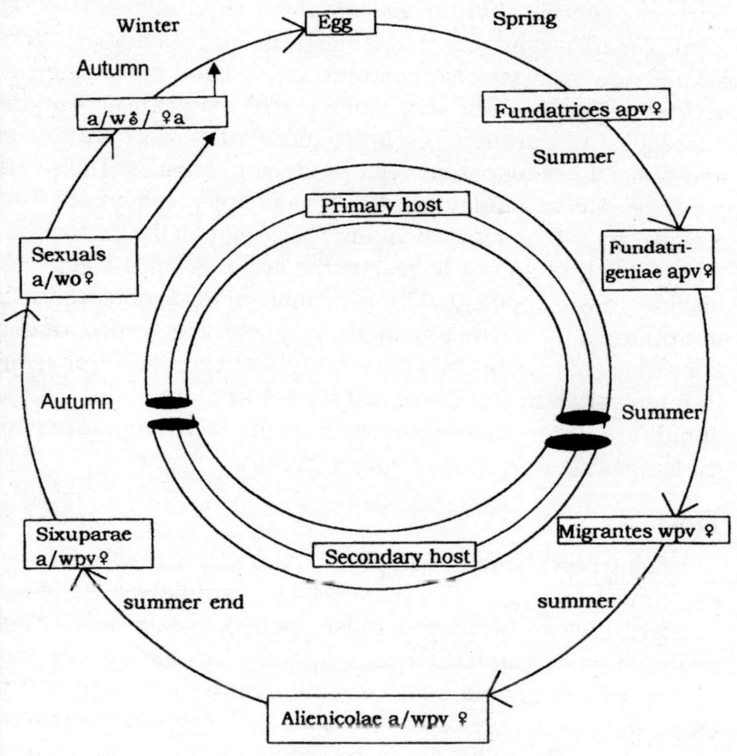

Figure 1b: Life History of an Obligatory Migratory Aphids

The complication in the life history occurs in the case of the obligatorily migratory aphids which have to undergo the following seven stages, egg, fundatrices, fundatrigenriae, migrants, alienicolae, sexuparae and sexuals. The diapausing eggs of winter hatch in spring into apterous, parthenogenetic viviparous females (fundatrices) which live on the primary host. After some generations, the fundarices produces fundatrigeniae which like their predecessor or apterous, parthenogenetic, viviparous females. The fundatrigeniae after 2-3 (or more) generations change to migrantes which are winged, parthenogenertic, viviparous females which through living on the primary host subsequently fly to the secondary host (of different species) to produce apterous and winged parthenogenetic viviparons females called alienicolae. The aliencolae after some generations and by the end of summer produce sexuparae which are also apterous and winged parthenogenetic viviparous females. The alatae (or winged)sexuparae now migrate once again to the primary host to produce sexuals which are apterous and winged but oviparous (egg producing) females. The sexuals produce apterous and winged males and apterous females. These females lay 3-4 (or some times one) large egg on the primary host which hatch in spring to restart the cycle. From the above life histories, it will appear that the predominant mode of reproduction in aphids is by parthenogenetic vivipary, the second ovipary appearing only briefly possibly to bring about genetic improvement through variation that the sexual reproduction provides. Tropical aphids are largely non-migratory a facultatively migratory while, the temperate ones are obligatory migratory.

Host

Being polyphagous, the aphids attack a large number of plants *e.g.* almost all vegetables like cabbage, cauliflower, Knol-knol, mustard, melons, beans, pea, pulses, potato, tomato, lady's finger, chilly, onion, cereals like wheat, sorghum, corn, barley, sugarcane, cotton, fruits like apple, peach, *Z. zizuba,* garden plants, plantations crops and so on, may be heteroecious and others autoecious and occasionally the same diversity occurs within a species at the subspecific level (Shaposhinikov, 1981).

There are no doubt that the natural enemies of aphids can reduce their rate of increase, occasionally dramatically and the use of hymenopterous parasitoids and predators in biological control of a few aphid pests would appear to have been successful (Sathe 2004b; Sathe and Bhoje, 2000; Sathe and Bhosale, 2001; Sathe and Patil 2003 etc.)

The known world fauna of aphids (Aphidoidea) consists of 4401 species, placed in 493 currently accepted genera (Blackman and Eastop, 1994), particularly from the subfamily Aphidinae, which are usually the important pests of agricultural and horticultural worldwide, damaging plants by sucking phloem sap and by transmitting viral diseases (Kennedy *et al.*, 1962; Canakcioglu, 1975; Blackman and Eastop, 1984; Conti, 1985). According to Raychaudhuri (1980), out of 379 species of aphids recorded from

North East Indian and Bhutan as many as 108 were associated with plants of economic importance, excluding those in the forest area. The aphids are mostly gregarious, in most cases the tender growing parts of the plants are infested. The infested plants shows following features

1. Curling of the tender leaves.
2. Twisting of the tender shoots and general devitalization of the plants.
3. Young infested seedlings succumbs.
4. Inflorescence may fail to open fully when infested.
5. Infested fruits fail to develop normally, may show malformation.
6. Twisting of pods.
7. Impaired development of healthy and viable seeds.
8. Affects subaerial part of plants *i.e.* roots resulting yellowing of foliages and stunting growth of plants.
9. Aphids form different types of galls on leaf and stem, those subsequently serve as temporary abodes for those aphids.
10. Aphids can cause damage to plants indirectly by their copious secretion of honey dew like substance which interfere with normal physiological processes like respiration and photosynthesis of the plants. Depositions of honey dew on the leaf surface allow forming sooty moulds which affect photosynthesis.
11. Aphids may cause nacrosis.
12. Aphids acts as vector for viral diseases. Out of about 300 known viral diseases of plants aphids can transmit as many as 90 diseases.

As Western Ghats is one of the hot spots of the world in biodiversity and aphids are quite destructive insects of various types of plants, the present study will add great relevance in understanding, conserving and protecting the biodiversity of both, host plants and aphids not only from forests but also from agro and horticultural ecosystems.

Chapter 3

Crop Aphids and their Control

See the Table 1.

Table 1

Sl.No.	Pest Species	Marks of Identification	Life Cycle	Damage	Hosts	Control
(I) Agricultural Crop Aphids						
A) Cereal Crops						
1.	Jowar aphid *Rhophalosiphum maidis* Ashmed	The jowar aphid is soft bodied, louse like, shows two pairs of transparent wings on thorax and one pair of cornicles on abdomen. It has 6 segmented antenna. Process terminalis is about 1.70 to 2.40. Abdominal dorsum is with blunt and fine hairs, Antennal segment 5th with 3 to 8 secondary rhinaria. Nymphs are similar with adults by shape but lacking wings and miniatures. Siphuncula, cauda and body length of aptera is 0.64 mm, 0.03 mm and 1.95 mm respectively	Aphids reproduce parthenogenetically 15 to 25 hours after the last nymphal moult. About 45 young ones are produced within 5 days. There are four instars in nymphal stage. Life cycle from egg to adult is completed within 15 or 16 days.	Nymphs as well as adults cause damage to crop by sucking the cell sap and affecting growth and yield of the crop. Specially they suck the cell sap from tender leaves, leaves turn yellow and plant remain stunted in severe infestation. The pest secrete honeydew like substance that cause sooty mould on leaf surface which affect photosyrthesis, growth and yield of the crop.	Jowar, Bajra, Maize, sugarcane and other cereal crops.	i) **Chemical**:Spray the crop with any of the following, 0.20% thiometon/ methyl demeton or 0.1 % carbaryl/ 0.03% malathion/ 0.03% Azadirachtin. ii) Natural control: A lady bird beetle: *Menochilus sexmaculatus* feed on aphids and suppress pest population. **iii)Preventive:** Collection and destruction of infested plant parts along with pest stages.

Contd...

Table 1–Contd...

Sl.No.	*Pest Species*	*Marks of Identification*	*Life Cycle*	*Damage*	*Hosts*	*Control*
2.	*Aphis adusta*	Soft bodied, louse like aphids	Reproduce partheno-genetically & viviparo-usly	Suck cell sap from tender parts of crop.	Jowar	i) **Chemical**: Spray: 0.02% phosphami-don or 0.03% mala-thion.
3.	The Barly aphid *Macrosiphum granarium* Kby.	It is green coloured and louse like. Nymphs are similar to adults by morphological features but they lack wings and are miniatures.	Life cycle is completed within 7-8 days from nymphs. The pest reproduce partheno-genetically females in winter. In summer winged males and females are produced. The winged forms migrate to various host plants like grasses. Females can reproduce viviparously by giving birth to young nymphs. In October aphids appear on wheat.	Both adults and nymphs suck the cell sap from tender leaves and ears. Infested leaves turn yellow which results in reducing the yield of the crop, form sooty moulds on leaves and affect photosynthesis.	Barley, wheat, millets, etc.	i) **Chemical** : spray- 0.03% malathion or Azadirachtin. 0.02 phosphamidon or 0.01% carbaryl. ii) **Preventive:** Collection & destruction of plants parts infested along with pest stage. iii)**Biological:** Use of natural enemies like Lady bird beetles.

Contd...

Table 1–Contd...

Sl.No.	Pest Species	Marks of Identification	Life Cycle	Damage	Hosts	Control
4.	The wheat aphid *Macrosiphum miscanthi* (Takahashi)	Winged and nonwinged forms are common in this species. The aphid is green coloured 3 mm in body length in aptera froms and 2.40 mm in alate forms. Siphunculus is 0.75 mm and 0.50 mm in aptera and alata respectively. Nymphs are similar to adults but smaller in size and lacking wings.	Reproduction takes place by parthenogenetically and viviparously. In cold weather pest multiplies very fastly and peaked in February–March. Approaching summer the winged forms are formed and migrated to the other plants and again appear on wheat in October–November.	As above	Wheat	As above
B) Sugarcane						
5.	The cane aphid *Melanaphis sacchari* Zeh.	The aphids is louse like, soft bodied with a pair of siphunculi on the back. Nymphs are nonwinged and smaller in size but similar to adults. Body length, siphunculi and cauda measures 1.53 mm & 1.61 mm, 0.09 mm & 0.12 mm and 0.15 mm & 0.12 mm in aptera and alata aphid respectively.	Reproduction is parthenogenetically and viviparously. Hence built their population in short time.	Both nymphs and adults suck the cell sap and inject toxins into plant as a result the juice becomes insipid and sugar recovery is badly affected. They also form sooty mould on leaves, affect photosynthesis, growth and finally the yield of crop.	*Rubia* sp., Sugarcane, *Paniculum* sp., *Sorghum vulgare*, *Pennisetum* spp.	**i) Chemical**: spray: 0.02% phosphamidon, 0.03% malathion or 0.03% Azadirachtin or 0.1% carbaryl. **ii) Natural enemies:** Lady bird beetles, lace wings. **iii) Preventive:** As above pest.

Contd...

Table 1–Contd...

Sl.No.	Pest Species	Marks of Identification	Life Cycle	Damage	Hosts	Control
6.	*Tetraneura javensis*	It is pear shaped aphid	As above	As above	Sugarcane	As above ii and iii
7.	The root aphid *Geocia lucifera* Z.	Adults are yellowish or pale white while, nymphs are yellowish but later they turn pale white. Pest is sporadic in appearance.	A single female can lay about 30-50 eggs in clusters on the undersurface of leaves. Incubation takes place within a week. The larval period is about 5 to7 days. Thus, the pest take 13-25 days for completion of life cycle from egg to adult. Many generations are completed in a year. The pest can reproduce partheno-genetically and viviparously.	Adults as well as nymphs suck the cell sap from roots of the plants and affect the growth and sugar formation adversely. The pest also secrete honeydew like substance on leaves and cause sooty moulds and affect growth and yield of the crop.	Sugarcane, Jowar, paddy	i) As above pest ii) 0.05% methy parathion or 0.01% malathion

Contd...

Table 1–Contd...

Sl.No.	Pest Species	Marks of Identification	Life Cycle	Damage	Hosts	Control
8.	Sugarcane wooly aphis. *Ceratovacuna lanigera* Zehnt, (figs. 184-195)	Adults are black coloured with two pairs of transparent wings. Forewings are larger than hindwings and about 2.70 mm and 1.20 mm in length and width respectively. Its body length is about 2.10 mm and 0.75 mm in width. There are four nymphal instars. They are yellowish or greenish yellow in colour and with a pair of siphunculi on 5th segment of abdomen. First instar nymphs are very active, wooly filaments are not seen on first two instars but in later instars they are densely present for their protection. Wooly fibers are quite soft, dry	It reproduce partheno-genetically, without males and vivi-parously. Nymphal stage lasts for about 16 to 17 days. The female can produce 3 to 5 nymphs per day with an average of 41 to 56.6 individuals in her life. Nymphal stage of wingless (apterous) form lasts for 23 to 32 days while in alate forms it is 32 to 40days. Winged forms migrate from one plant to another and field to field. Wingless aphid produce 60 eggs and winged 10 eggs. Sexuparae generation is also noticed in the pest.	Both adults and nymphs suck the cell sap from leaves through stomata and utilize amino acids and sugar for their own life activities. They secrete honeydew like substance which cause sooty mould on the leaves, affect photosynthesis, growth of crop and yield of the crop. Sugar recovery is badly affected, juice becomes insipid.	Bamboo, *Miscanthus sinensis, Cyanodon dactyiedon*	**i) Preventive:** Collection & destruction of infected plant parts along with pest stages. ii) Sugarcane is cultivated by ridges and furrow method or by bed method. iii) Infested canes should be avoided for plantations (seeds). After selection of seeds, left over leaves should be burnt. iv) Seeds should be treated with 300 ml malathion + 100 ml water for 15 minutes. v) Use balanced chemical fertilizers. Excess use of nitrogenous

Contd...

Table 1–Contd...

Sl.No.	*Pest Species*	*Marks of Identification*	*Life Cycle*	*Damage*	*Hosts*	*Control*
		and hydrophobic formed by amino acids and nitrogenous material. Both nymphs and adults are congregated on cane leaves with huge number appearing as a cottony mass.				fertilizers promotes growth of aphids. **Chemical control:** Spray the crop with any of the following insecticides- metasystox 0.0375%, endosulphan 0.05%, rogor 0.045%, malathion 0.1%, Acephate 0.1% or dusting of 4% malathion 40 kg./ha. Soil application of phorate 10 to 15 kg/ha to 20 kg/ha for medium and full grown crop respectively. **Biological control:** Release following BCAS. *Conobathra aphidovora* - 1000 larvae/ cocoons/ha.

Contd...

Table 1–Contd...

Sl.No.	Pest Species	Marks of Identification	Life Cycle	Damage	Hosts	Control
						Micromus sp. 2500 larvae or cocoons/ ha. Syrphid fly *Eupeodes canfrator*-1000 larvae/ cocoons/ha. Lacewing *Cryso-perla carnea*- 2500 eggs or larvae/ ha. (No adults are released).
C) Oilseed Crops						
9.	The groundnut aphid *Aphis craccivora* Koch. Fig. 18-23 & 27)	Aphis are greenish black, winged aphids are with black wings. Both winged and wingless forms are produced by this pest. Apterae measures for 2.11 mm, 1.46 mm, 0.42 mm and 0.25 mm in body length, antenna, siphunculus and cauda respectively. While	Reproduction is parthenogenetically and viviparously. Winged forms can produce wingless forms. Female produce 8-30 young ones without fertilization within 10-12 days of her life span. There are 4 instars in the nymphs. Nymphal period is 5-8	Both adults and nymphs suck the cell sap from tender leaves and affect the growth and pod setting. The aphid found sitting from underside of leaf and suck the cell sap. They cause stunting in plants. Flowers drop down, form sooty moulds on leaves and	It is polyphagous pest. It infests members of 18 families of plants. Ex. peas, pulses, cereals, fiber crops, oil seed crops, beans, etc.	**i) Chemical:** Spray the crop with any of the following pesticides 0.08% thiometon, 0.04% monocrotophos, 0.03% meta systox, Dusting: 3% morphothion, 5% malathion. **ii) Preventive**: as above pest.

Contd...

Table 1–Contd...

Sl.No.	Pest Species	Marks of Identification	Life Cycle	Damage	Hosts	Control
		alata measures 1.38 mm, 1.29 mm, 0.12 mm and 0.10 mm in body length, antenna, siphunculus and cauda respectively. Young nymphs are brownish.	days. Adults starts giving birth within 24 hours. Breeding occurs throughout the year.	affect the yield of crop. It also acts as vector for viral diseases.		iii) **Biocontrol:** Use of *Menochilus sexmaculatus* as predator and Syriphid, *Ishiodon javana*
10.	The mustard aphid *Lipaphis erysimi* Kalt. Fig. 86-91 & 97)	Adult is pale greenish which measures about 1.52 mm, 0.18 mm and 0.15 mm in body length, siphunculus and cauda respectively. It is louse like. Nymphs are similar with adults by forms but smaller in size. Both winged and nonwinged forms are common.	Winged forms are emerged in autumn and spring and migrate to various crops for feeding purpose. The population is peaked from December to March since they feed on various cruciferous oilseed and vegetable crops. They migrate to mountain or hilly region during summer. The pest breeds parthenogenetically and viviparously. A female gives birth to	Both nymphs and adults suck the cell sap from tender parts of the crop specially leaves, stem, inflorescences and developing pods, as a result vitality of crop is reduced. Flowering and fruiting bodies drop down. Pod setting is adversely affected and the yield of crop too.	Brassica vegetables, raya, tarmira, mustard, toria etc.	i) **Preventive:** Collection & destruction of infested plant parts along with pest stages. ii) **Chemical:** spray 0.02% phosphamidon or 0.05 % Roger or dust- 5% malathion. iii) **Biological:** Following parasitoids cause mortalities in pest– *Ischiodon scutellaris* ii–*Syrphus*

Contd...

Table 1–Contd...

Sl.No.	Pest Species	Marks of Identification	Life Cycle	Damage	Hosts	Control
			about 25-130 nymphs. Nymphal period is 7-10 days. During cold and cloudy condition it multiplies very fastly.			*serarius*. Predators- *Crysopa scelestes*, *Coccinella septumpunctata*, *Menochilus sexmaculatus*. Parasitoids: *Diaeretiella rapae Lipolexis gracilus* Parasitize pest-Fungus: *Entomopthora coronata*, *Cephalophorium aphidicola* kill pest.
11.	The Green aphid- *Myzus persicae* (Sulzer) (fig. 119- 128 & 132)	Yellow green winged and nonwinged forms are common	Life cycle described under vegetable crops.	Damage mustard plant by sucking cell sap and affecting the yield of crop.	Mustard, peach, bean, potato, cruciferous crops.	i) **Preventive:** As above ii) **Chemical:** Spray: 0.02% phosphamidon.

Contd...

Table 1–Contd...

Sl.No.	*Pest Species*	*Marks of Identification*	*Life Cycle*	*Damage*	*Hosts*	*Control*
12.	The Safflower aphid *Dactynotus carthami* H.R. L.	Adults and nymphs are black, soft bodied, louse like, winged and nonwinged forms are common. Head is smooth, antenna 6 segmented, flagellum imbricated, primary rhinaria ciliated. Rostrum reaches upto hindcoxae and nymphal hindtibiae smooth.	It reproduces parthenogenetically and viviparously.	Pest colonies are located on various tender parts of the crop which suck the cell sap and affect the yield of the crop. The pest also secretes honeydew like substance which cause sooty moulds and affect photosynthesis and yield of the crop adversely.	Safflower	i) **Preventive:** As above pest ii) **Chemical:** As above iii) **Biological:** Lady bird beetles feeds on aphids.
D) Pulses						
13.	The Tur aphid *Aphis medicaginis* Kle.	Aphids are black coloured. Head is without any lateral frontal tubercles. Antenna is 6 segmented. Apterae forms without secondary rhinaria. Cauda is elongated. Siphunculi bearing 4-12 hairs. Midthoracic furca without a stalk. Apterous forms are bluish greenish and with white wax.	Reproduction viviparously and parthenogenetically. It is monoecious holocyclic.	Both nymphs and adults suck the cell sap from tender parts of pigeon pea crop. Specially leaves, flowers and developing pods etc.	*Cajanus cajan*, Lucerne, *Aster*, *Cosmes*, *Polygonum*, *Rumex*, *Portulaca*, etc.	i) **Preventive:** As above pest ii) **Chemical:** As above pest iii) **Biological:** Lacewings and some lady bird beetles feed on aphid populations.

Contd...

Table 1–Contd...

Sl.No.	*Pest Species*	*Marks of Identification*	*Life Cycle*	*Damage*	*Hosts*	*Control*
14.	The pea aphid *Macrosiphum pisi* Kalt.	Aphids are soft bodied louse like with two bars on abdomen. Both nymph and adults are similar except the nymphs are smaller in size and without wings. Bars are cylindrical, legs are smooth and cauda is elongated.	Reproduce Partheno-genetically and viviparously.	Nymphs and adults suck the cell sap from tender parts of crops such as leaves, flowers and developing pods. They create sooty moulds, affect photosynthesis and yield of the crops.	Pea, cow pea	i) **Preventive:** As above pest ii) **Chemical:** As above pest iii) **Biological:** As above pest.
15.	The groundnut aphid *Aphis craccivora* Koch.	See under groundnut	See under groundnut	Nymphs and adults suck the cell sap from tender leaves, flower and developing pods of mung, udid, cowpea, beans, wals, etc. and affects the growth and yield of crop.	Mung, Udid, Cow pea, Wals, Beans, etc.	As above pest.
16.	The pulse aphid- *Therioaphis sexmaculatus* Buck.	Aphids are pale yellowish, soft bodied with knobbed cauda. They measures about 1.85 mm in body length.	Reproduce partheno-genetically and ovoviviparously.	As in above crops.	Mung, Udid, Cow pea, Wals, Beans, etc.	As above pest

Contd...

Table 1–Contd...

Sl.No.	Pest Species	Marks of Identification	Life Cycle	Damage	Hosts	Control
E) Fiber Crops						
17.	The cotton aphid *Aphis gossypii* Glover (Fig. 28- 34 & 40)	Adults are soft bodied and greenish brown in colour. Winged and nonwinged forms are common. Alate form measures for 1.53 mm, 0.30 mm and 0.15 mm in body length, siphunculus and cauda respectively while apterae forms measures for 1.48 mm, 0.18 mm and 0.16 mm in body length, siphunculus and cauda respectively. Nymphs are similar to adults by shape but lacking wings.	The pest reproduce parthenogenetically and viviparously, giving birth to 8- 20 nymphs per day per female. The nymph moults four times to become adult. Nymphal period is 7–10 days. In dry climate the pest multiplies fastly.	Adults as well as nymphs cause damage to cotton crop by sucking cell sap. As a result, leaves become curly, they turn yellow, dry and dropdown and affect the growth and yield of the crop. Flowering and fruiting bodies of cotton dried and drop down, bolls open pre-maturely producing lint of poor quality, whose market value is adversely affected. The pest also secrete honeydew like substance which form sooty moulds on leaves and affect the photosynthesis and crop yield adversely. Younger plants are attacked more than older ones.	Cotton, Chinarose, *Cassic glauca, Tecoma capensis, Rosa* sp., potato. It is highly polyphagous found on about 62 families of plants.	**i) Preventive:** Collection and destruction of infested plant parts along with pest stages. **ii) Chemical:** Spray crop with any of the following. 02% phosphamidon, or 0.2% methyl demeton, or roger, dichlorvos or monocrotophos. **iii) Biological:** Lady bird beetles and lacewings feed upon the pest stages and cause mortalities.

Contd...

Table 1–Contd...

Sl.No.	Pest Species	Marks of Identification	Life Cycle	Damage	Hosts	Control
18.	The Ambadi Aphid *Aphis malvae* Koch.	The aphid is louse like with faint legs, elongated, cauda shorter than bars bearing hairs, wing venation normal with brown boarder.	Reproduce parthenogenetically and viviparously	Pest suck cell sap from the tender portions of the plant (Ambadi), which is fiber crop of Maharashtra. It forms sooty moulds over leaf surface, affect photosynthesis, growth and yield of crop adversely.	Ambadi, *Hibiscus* sp, *Cucurbits,* Sweet potato etc.	As above pest **Chemical:** Spray: i) 0.02% phosphamidon (or) malathion or Azadirachtin
F) Vegetable Crops						
19.	The green peach aphid *Myzus persicae* Sulzer (Mustard aphid)	Winged aphids are pear shaped, yellowish green and measures about 2.25 mm in body length. Siphunculus and cauda measures about 0.40 mm and 0.20 mm respectively. Apterous forms measures about 1.9 mm, 0.50 mm and 0.20 mm in body length,	Aphids pass the winter in egg stage. The eggs are laid on host plants and hatch out in spring. They grow as an adult within 5 to 7days. Then the adults give birth to wingless forms. Through siphunculus they produce pheromone which signals them against danger.	Aphids found on the mustard with large populations sucking cell sap from tender leaves, flowers and developing pods. They affect growth, photosynthesis and yield of crop. They suck cell sap of potato crop and secrete honeydew like	Mustard, cruciferous plants, potato, bean, peach, radish, tobacco, etc.	i) As above pest **ii) Chemical:** Spray 0.03% malathion, or 0.02% dizinon or 0.03 Azardirachtin.

Contd...

Table 1–Contd...

Sl.No.	*Pest Species*	*Marks of Identification*	*Life Cycle*	*Damage*	*Hosts*	*Control*
		siphunculus and cauda respectively. Nymphs are similar to adults by shape but they are smaller and without wings. Eggs are black and shining.	The pheromone is called as "Alarm pheromone"	substance which create sooty moulds on leaves further, it affect photosynthesis, growth of crop and yield adversely. The aphid can transmit viral diseases to plants.		
20.	The cotton aphid *Aphis gossypii* Glover (Potato aphid)	Given under fiber crops	See under fiber crops	Both nymphs and adults suck cell sap from potato leaves and vines as a result tuber setting is adversely affected. They also responsible for sooty moulds on leaves which affect photosynthesis and growth of crop and finally tuber setting.	Potato, mustard, peas, cotton, Brinjal, chilies, banana, okra, etc.	As above.
21.	The potato aphid. *Aphis malvae* Koch.	Given under fiber crops	As above	Both nymphs and adults suck the cell sap	Sweet potato, cucurbits, etc.	As above
	Aphis brassicae	Louse like aphid	Suck cell sap	As above	Radish	As above

Contd...

Table 1–Contd...

Sl.No.	Pest Species	Marks of Identification	Life Cycle	Damage	Hosts	Control
22.	The Cabbage aphid *Brevicoryne brassicae* Linn.	Apterae forms are medium sized,(1.6-2.6mm) long, grayish green or dull mid - green. Its body is covered with grayish-white mealy wax. Alatae are with dark head & thorax, Siphunculi black on abdomen. Winged forms measures from 1.6 mm to 2.8 mm in body length.	It is monoecious holocyclic in colder regions & anholocyclic where winter climate is mild. Reproduce parthenogenetically & viviparously. There are four nymphal instars in the pest.	Both nymphs & adults suck the cell sap from cabbage & cauliflower & affect quality of human food. They suck cell sap from leaves, cabbage bulb & leaves & flowers of cauliflower. The pest transmit certain diseases, It is victor of about 20 plant viruses Ex. cabbage blac kings spot, cabbage ring necrosis, cauliflower mosaic etc.	Many cruciferae crops specially cabbage, cauliflower, radish, swede, mustard, Brussels sprout, etc.	**i) Preventive:** collection & destruction of infested plant parts along with pest stages. **ii) Chemical:** spray: 0.03% malathion or 0,03% Azadirachtin. **iii) Biological:** Use Hymenopterous parasitoid *Diaretialla rapae* against this pest.
(II)	**Horticultural Crops**					
23.	The apple wooly aphid *Eriosoma lanigerum* (Hausmann)	Both winged and nonwinged forms are common in this species. Wingless forms are purple/red/brown in colour but covered with thick	It is anho ocyclic on apple throughout the world. The pest reproduce by parthenogenetically females only. Each female gives 116	Both nymphs and adults are destructive to apple crop. They suck the cell sap from bark of twigs when they are terrestrial and from roots when they are	Apple	**I) Preventive:** i) Plougging and digging is essential for exposing undergrqund populations

Contd...

Table 1–Contd...

Sl.No.	Pest Species	Marks of Identification	Life Cycle	Damage	Hosts	Control
		white flocculent waxy component. They are 1.2 to 2.6 mm in body length. Winged forms are with reddish brown abdomen covered with waxy component and measures from 1.8 mm to 2.30 mm in body length.	young ones. The nymph moults for 4 times to become adult The nymphal period during January-February is 35 to 42 days and 30 days in August- November while during April July it is about 11 to 20 days Its population is actively multiplied from the months, March to September. In July, September winged forms appear. They feed on terrestrial parts of apple but, migrate to roots, underground part of the crop in December. While in May the pest again migrate to terrestrial parts of the crop for feeding purpose.	underground as a result knots appear on the roots which affects the growth of plant. The infested twigs shrivel and dies affecting the yield of crop seriously		ii) Collection and destruction of aerial infested parts of crop along with pest stages **II) Chemical:** For aerial infestation spray the crop with any of the following- 0.05% carbaryl/ 0.03% ethyl parathion or 0.05% diazinon. Summer treatment of aerial parts: Spray 0.04% nicotine sulphate or 0.08% malathion or 0.08% Azadirachtin. Treatment to roots: 0.05% methyl demeton or fumigation with Para dichlorobenzene in spring 15 cm deep in soil.

Contd...

Table 1–Contd...

Sl.No.	Pest Species	Marks of Identification	Life Cycle	Damage	Hosts	Control
						III) Biological: Use Hymenopterous parasitoid, *Aphelinus mali* which kills the pest by parasitization. IV) Use of pest resistant varieties of apple: ex. Merton type No 778 & Malling type - XIII, are useful.
24.	**The peach leaf curl aphid:** ***Brachycaudus helichrysi*** **Kalfenback**	It is louse like and yellowish with dark stripes on the head. Its colour is changed with respect to host plants. On peach it is light green while on golden rod it is greenish yellow. Apterae measures from 0.9 mm to 2 mm and alata from 1.1 mm to 2.2 mm in body length.	On golden rod it is active from June to October while from February to March it is active on temperate fruit trees. In winter they are in egg stage. Eggs hatch in spring and newly emerged nymphs starts sucking cell sap of primordial leaves of host plants. Non winged nymphs become adults	The pest suck the cell sap from tender leaves of peach as a result they become curly. Leaves get twisted in severe infestation which affect fruit setting adversely. The pest also transmit viral diseases such as plum pox, cucumber mosaic, cineraria mosaic, potato virus, etc.	Peach, golden rod, plum, cucumber, cineraria, potato, etc.	**I) Preventive:** Collection and destruction of crop parts infested along with pest stages. **II)Chemical:** Spray the crop with diesel oil emulsion mixed with Bordeaux mixture, or 0.1 % carbaryl, or 0.08% malathion, or 0.03% DDVp.

Contd...

Table 1–Contd...

Sl.No.	Pest Species	Marks of Identification	Life Cycle	Damage	Hosts	Control
			within four weeks. No males are produced in this generation. Thus, the female produce parthenogenetically only. Females larviposite about 50 individuals within her life span of about 13 days. On this fruit plant 3-4 generations are completed. Winged males and females are produced in warming seasons. Such winged migrates to alternative host plants like golden rod and produce sexuals. 4 to 5 generations are completed on golden rod crop. In early November winged females are produced once again. They migrate back to peach or other fruit plants for			

Contd...

Table 1–Contd...

Sl.No.	Pest Species	Marks of Identification	Life Cycle	Damage	Hosts	Control
			laying eggs on fresh material of the crop. Oviposition is completed in mid December finally the female dies.			
25.	The Banana aphid *Pentalonia nigronervosa* Coq.	Aphids are oval with reddish brown to black colour. Both winged and nonwinged forms measures from 1.1 to 1.8 mm in body length. Antennal base and apex is dark and rest of the antennal part is pale coloured. Siphunculi and cauda measures for 0.34 mm and 0.12 mm respectively in apterae. Antenna is as long as body, siphunculi are slightly clavate and long.	It is completely anholocyclic form, sexual forms are also noticed. The pest reproduce parthenogenetically and viviparously. The pest found in India and Nepal on banana crop.	Both nymphs and adults suck the cell sap from tender leaves of banana. They are found sitting in the developing whorl of the leaf and sucking cell sap. They secrete honeydew like substance and create sooty mould on leaf surface which affect photosynthesis, growth of the plant and fruit setting adversely. It also acts as vector of banana bundy top, banana mosaic and chlorosis, heart rot, katte mosaic, etc.	Banana, Abaca, *Heliconia*, taro, *Caladium*, *Zingiber*, *Musa spp.*, *Curcumba longa*, etc.	I) **Preventive:** collection and destruction of infested leaves along with pest stages. II) **Chemical:** spray 0.03% phosphamidon, or 0.05% DDVP or carbofuron/phorate/ disulphoton at 4.5 kg/ha and 2.25 kg/ha

Contd...

Table 1–Contd...

Sl.No.	Pest Species	Marks of Identification	Life Cycle	Damage	Hosts	Control
26.	The citrus aphid - *Toxoptera citricidus* Kirkaldy	Adults are very dark brown to black, shiny and about 1.5 to 2.4mm in winged & nonwinged forms. Antennae are with much longer hairs. Nymphs are brownish in colour.	It is totally anholocyclic form and an abortive holocyclic on citrus. It reproduces parthenogenetically and viviparously.	Both nymph and adult suck the cell sap from tender leaves and stems of citrus plant as a result flowers drop down, leaves become curly, turn yellow and drop down and fruit setting is adversely affected. They secrete honeydew like substance on leaves and form moulds which affects photosynthesis, growth and yield of the crop adversely. They also act as vector for "citrus tristeza virus" "citrus vein enation". Stem pitting virus and "bud union decline" virus of lemon, etc.	The pest attacks 21 families of plants in addition to citrus and widely distributed in the world.	**I)-Preventive** Collection and destruction of infested plant parts along with pest stages. **II)-Chemical:** 0.03% phosphamidon or 0.05% DDVP or 0.05% dizinon or 0.05% Azadirachtin spray

Contd...

Table 1–Contd...

Sl.No.	*Pest Species*	*Marks of Identification*	*Life Cycle*	*Damage*	*Hosts*	*Control*
27.	*Toxoptera auranti* (B) (Black citrus aphid)	Black coloured	Given under mango	Suck cell sap	Coffee, fig, citrus mango, Anona, etc	As above
28.	The Mango aphid *Toxoptera auranti* (B) (Tamarid aphid)	Aphids are oval in shape and soft boded. They are reddish brown or black in body coloration. Antennae are with white & black bands. Siphunculi and cauda are black in colour. Nymphal forms are brownish. Winged forms are with brown abdomen.	It is entirely anholocyclic. No sexual morphs are seen in their any generation of the year. It reproduce parthenogenetically and viviparously.	Both nymphs and adults suck the cell sap from inflorescence and tender leaves as a resut the flowering and fruiting bodies of mango drop down and make plant completely sterile in severe infestation. It is very bad pest of mango especially Hapus varies.	It attacks about 120 host plants in addition to mango, citrus, coffee, tea, tamarid, Anona and ficus.	**I) Preventive** - collection and destruction of infested plant parts along with pest population **II) Chemical:** Spray-0.03% malathion, or 0.05% DDVP or 0.05% Azadirachtin

Contd...

Table 1–Contd...

Sl.No.	Pest Species	Marks of Identification	Life Cycle	Damage	Hosts	Control
(II)	**Ornamental Crops**					
29.	The Rose aphid - *Macrosiphum rosae* (L)	Both winged and wingless forms are common. Winged aphids are with very distinct black sclerites on abdomen sides and measures from 2.2mm to 3.4mm in body length. Wingless forms are broadly spindle shaped with mid to dark green or deep pink to red brown or magenta coloured. Its siphunculi and head are black and cauda yellow.	It is heteroecious holocyclic between Rose and Valerianaceae or Dipsacaceae in temperate regions. Reproduce parthenogenetically and viviparously.	Aphids suck cell sap from tender portion of Rose plant and affect the growth adversely, thereby formation of flowers too.	Rose, some members of families Valerianaceae and Dipsacaceae.	As above
30.	The rose aphid *Macrosiphum rosaeiformis* Das	Louse like aphid have soft body and a pair of bars on the abdomen. Winged and nonwinged forms are common.	Reproduce Parthenogenetically and viviparously.	Suck cell sap from tender parts of rose and affects growth and formation of flowers on the plant.	Rose	I. As above II. Ladybird beetles feed on aphids.

Chapter 4

New Aphids on Crops and their Control

Introduction

In the animal Kingdom insects form the largest group. They are divided into 29 orders that are in turn broadly grouped into wingless (ex. silverfish) and winged (ex. bugs, butterflies, moths, beetles, flies etc.) forms. Insects are the only invertebrates which can fly. They seem to be have made their appearance around 360 million years ago during a great insect 'boom' in the bush, gloomy and fern forests of the world. Insects are of the most successful groups of animals, as they have been around much before dinosours appears on the earth. Significantly, they have been managed to survive the on slaughts of *Homo sapien*.

According to Cheeliah and Ganathilagaraj (1993), about ½ million species have been recorded in the scientific literature and several million insects are awaiting their discoveries mainly in tropics. India is considered to be one of the richest centre of biodiversity in the world. Indian insect biodiversity consists of more than 60,000 species described (Ananthakrishnan,1993; Gupta 2004). Taxonomy has never been an attractive profession, firstly, it is regarded as unrewarding, secondly, it is given step motherly treatment from financial agencies and thirdly, the student find more glorious fields like molecular biology and biotechnology. Presently, the picture is quiet different and fast changing in favour of taxonomy and biodiversity. Taxonomists are presently needed in several institutes such as Universities, Research Institutes, Museums, and Central and State Govt. agencies, Industries, Zoos, etc. Taxonomists play very important role in the fields of applied biology including Public health, pest management, environmental problems, national defense, wild life management, forensic science etc in identifying the species.

Biodiversity play very important role in sustainable development of a region or a country. Therefore, in the present study biodiversity of aphids have been studied with respect to taxonomy.

Evolutionary Account

As like Adelgids, phylloxerids, aphids also belong to superfamily Aphidoidea of suborder Homoptera. All these insects are cell sap sucking insects and small in size, either winged or wingless, if present, are membranous. Although the oldest fossil aphid known, *Triassoaphids cubitus*, is from the triassic period, it is likely that the Aphidoidea evolved 280 million year ago, in the Carboniferous, from the same stock that gave rise to the new extinct Archescytinidae and Permaphidopsidae, which are thought to have lived on primitive gymnosperms like, the Cordiatales and Cycadophyta. Reproduction by means of unfertilized eggs is common to all three families of the Aphidoidea, and as the adelgids and phylloxerids appeared in the late Carboniferous or early Permian, parthenogenesis possibly evolved before these three families separated, over 200 million years ago. The resultant cyclical parthenogenesis, in which periods of parthenogenetic reproduction alternate with sexual reproduction to give a holocycle, is thought to have evolved in a seasonal climate, possibly that associated with the glacial period in the lower Permian. However, viviparity, another characteristic feature of aphids, must evolved later as the Archescytinidae and the modern adelgids and phylloxerids are oviparous (Heie, 1967). The characteristic shape and veining of their wings and the structure of their proboscis and legs, had evolved by the Jurassic, whereas the cauda and siphunculi appeared later in the cretaceous (Shaposhnikov, 1977).

Primeval aphids were probably polyphagous and fed on the parenchymatous and phloem tissue of plants. That the Aphidoidea are a quarter of the size of the Archecytinidae and Permaphidopsidae. As most of the extent species of all three families are monophagous, it could be argued that monophagy, like parthenogenesis, evolved early in the history of the Aphidoidea. The simple nymphal eye of 3 lenses, the triommatidium, is a characteristic which aphids share with adelgids and phylloxerids and is also associated with the reduction in body size that came with the evolution of parasitic way of life.

The Aphidinae, the largest sub-family of modern aphids, is not represented in the fossil record until the late tertiary and is associated with the evolution of the angiosperms, in particular the rosaceae. Tribes of second largest subfamily, the Drepanosiphinae, developed much earlier in the upper cretaceous and early tertiary. There are no paleontological evidences of the evolution of the third largest subfamily, the Lachninae. The lachnids have genetically been regarded as primitive. However, two genera, *Aphis* in the Aphidinae and *Cinara* in the Lachninae, each have a large number of species that are difficult to distinguish from one another. This and the absence of fossil lachnids led Heie (1967) to suggest that the Lachninae are of recent origin.

HISTORICAL REVIEW OF WORK

The important work on aphids from different parts of the world refers to Van der Goot (1913-1917)–from East Indies, Theobald (1915) from Africa, Mordvilko (1914, 1919) from USSR, Das (1918) from Lahore (Pakistan), Takahashi (1919-1965) and others from various South East Asian countries, Hille Ris Lambers (1938-1953) from Europe, Palmer (1952) from Rocky Mountain region, U.S.A., Cottier (1953) from New Zealand, Bodenheimer and Swirsky (1957) from Middle East, Eastop (1958) from East Africa, Heinze (1961) from Central Asia, Shaposhnikov (1964) from USSR and Stroyan (1964) from Great Britain.

Aphids of India have been studied by Barlow (1890), Buckton (1893-1898), Van der Goot (1916-1917), George (1925-1928),

Krishnamurthy (1928), Szelegirwicz (1969) and Smith (1972). Since then a sizable number of papers on Indian aphids have been published.

Basu (1961a, 1961b, 1964, 1967, 1969a and 1969b) described some new genera, species and subspecies of aphids from Darjeeling district of West Bengal. Basu and Banerjee (1958) studied aphids on economic plants of West Bengal. Basu and Ris (1968) described one new genus and three new species of Indian aphids. Basu *et al.* (1970, 1972, 1973, 1974) studied five new species and thirty new records from Asam and one genus *Betacallis* from North East India.

Chakrabarti *et al.* (1970, 1971, 1972a, 1972b) described new genera, species and new records from Kumaun hills of Himalaya. Chakrabarti and Raychaudhuri (1974, 1978a, 1978b) studied new subgenus from India and North East India and Bhutan. Chatterjee *et al.* (1961) described the aphids of Manipur State.

David (1958a, 1958b, 1969, 1972, 1975) described new genus and three new species from South India. David *et al.* (1970, 1971, 1968, 1970, 1972) reported new genera, species and new morphs from North Western India. David and Rajasingh (1969) studied new records from Assam. Ganguli and Ghosh (1965) studied aphids of economic importance in Tripura.

Ghosh (1969, 1973, 1974a, 1974b, 1974c) reported new species and new records of aphids from the North West Bengal and adjacent countries. Ghosh *et al.* (1971a, 1971b, 1971c, 1971d, 1969, 1970) reported new genus, new species and new records from different parts of Eastern India and Bhutan. Ghosh *et al.* (1970, 1971a, 1971b, 1972) studied aphids from eastern India. Ghosh and Raychaudhuri (1959, 1962a, 1962b, 1968a, 1968b, 1968c, 1969, 1971, 1972a, 1972b, 1973) reported new records of aphids from Calcutta, Assam, Sikkim and Eastern India.

Ghosh (1969, 1970a, 1970b, 1972a, 1972b) studied new species and subspecies from Uttar Pradesh and North-East India. Ghosh *et al.* (1976, 1970, 1971a, 1971b) described aphids from eastern India. Lal and Siddiqui (1952) studied biology of the peach leaf curling aphid from Kumaun Hills.

Raychaudhuri (1956, 1973, 1974) studied taxonomy of the aphids, revision of Greenidea from Eastern Himalaya. Raychaudhuri and Banerjee (1974) reported the genus *Melanaphis* with descriptions of some new taxa from India; Raychaudhuri and Chatterjee (1977) described two new species and a new morph of Greenideinae from West Bengal; Raychaudhuri and Ghosh (1958) studied aphids of Calcutta with special reference to the newly recorded host plant families; Raychaudhuri and Pal (1974) contributed on a new genus *Subtakecallis* and a new species *Subtakecallis brevisetosus* and Raychaudhuri *et al.* (1973) described one new genus and two new subgenera and fifteen new species of Greenideini from Eastern India. Verma (1965, 1969) studied additional aphid fauna and described a new genus, some new species and new records of aphids from North West India.

George (1927) reported South Indian Aphididae; Goot (1916, 1917) studied some undescribed aphids from collection of Indian Museum and noted some Indian aphids; Stary and Ghosh (1978) studied aphid parasitoids from Meghalaya.

Gulmamullah (1940) recorded aphid fauna of Delhi; Krishnamurthy (1928) studied Aphididae of Mysore; Kumar and Lavinge (1970) described a new genus and new species of Callipterinae from India; Menon *et al,* (1968) reported taxonomic studies of Indian Aphididae and aphid fauna of Delhi; Quednau and Chakrabarti (1976) described a new genus and three new species of aphids from North West India.

Behura (1963) studied aphids of India; Bhattacharya *et al.* (1980) noted aphids of North West India; Buckton (1893, 1896) described new gall making aphid from North Western Himalayan region; Fotedar and Kapur (1943) studied first record of the sexual forms and oviparous reproduction of woolly aphid from Kashmir.

Maity (1980) studied aphids of North West India with special reference to Garhwal, Himalaya; Mondal (1977) reported aphids of Sikkim; Pal (1975) studied aphids of Graminaceous plants in North Eastern Himalaya.

Agarwala (1980) studied some species of aphids in Sikkim and Bhutan; Agarwala and Ghosh (1984) reported checklist of Aphidoidea of India; Agarwala and Ghosh (1985) also studied biogeographical consideration of Indian Aphididae; Agarwala and Martin (1993) contributed on the genus *Anomalosiphum*; Agarwala *et al.* (1982) described one new subgenus and three new species and one new subspecies from Sikkim.

Raychaudhuri (1980) contributed on aphids of North East India and Bhutan and Ghosh (1980) on the fauna of aphids of sub family Chaitophorinae from India and adjacent countries. Similarly, Ghosh (1984) contributed on subfamily Pemphiginae of aphids and Ghosh (1982) on fauna of aphids of subfamily Lachninae. He (1988) also contributed on the aphids of sub families Pholemyzinae, Anoecinae and Hormaphidinae. Ghosh and Quednau (1990) studied the fauna of subfamily Drepanosiphinae and Ghosh and Agarwala (1993) the aphid fauna of subfamily Greenidinae. Recently, Sathe (1992) and Jadhav and Sathe (2006 a, b) recorded 40 species from Western Ghats and Maharashtra.

MORPHOLOGICAL CONSIDERATION

The body of aphid (Figures 2 and 3) is divisible into the head, thorax and abdomen, but divisions are sometimes not found clear in certain species.

Head (Figure 4)

The head (Figure 4) is dorsoventrally flattened, with clypeus fixed on ventral side and front on dorsal side which is flat. Fron may be undulated or flat, sinuate etc. The lateral frontal tubercles of head are if present, parallel, diverging and converging type. The dorsal surface of head may be smooth, scabrous and spinulose type. The dorsal cephalic hairs are of different types, the wax pores sometimes wax plates also present on head. Head consists of paired antenna (Figure 5) placed in antennal sockets. The antenna may vary with the segments (1 to 6). Antenna consists scape, pedicel and flagellum. The scape and pedicel are smooth, rugose, spinulose etc, the flagellum may also be smooth or may not bear spinules. Hairs on flagellum may be sparse or dense and flagellar surface may be smooth or imbricated. The last segment of flagellum has a stout basal portion, the base and short to very long slender terminal portion, the processus terminalis. The primary rhinarium is placed at the junction of the base and processus terminalis and few small accessory found adjacent to it. It may be round, oval, star shaped, elongate etc. The secondary rhinaria are always variably present in alatae and some apterous morphs of different flagellar segments. It may be round, elongate, oval subanal etc; both rhinaria may be ciliated or non ciliated, may or may not be protuberant.

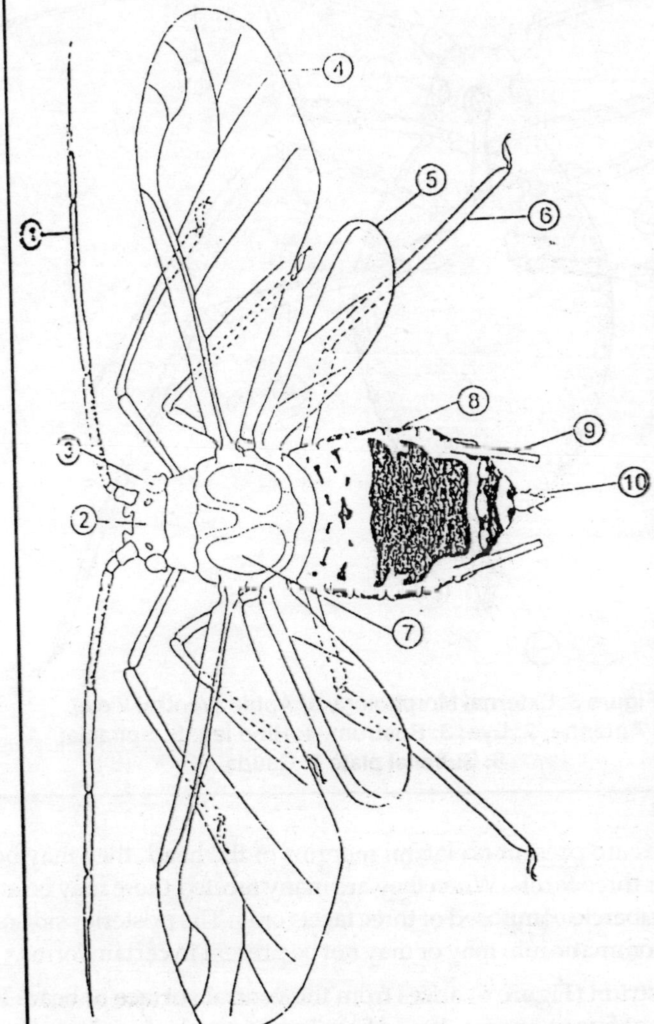

Figure 2: External Morphology of Aphid (Dorsal View)
1: Antenna; 2: Head; 3: Eye; 4: Fore wing; 5: Hind wing; 6: Hind leg; 7: Thorax; 8: Abdomen; 9: Siphunculus; 10: Cauda

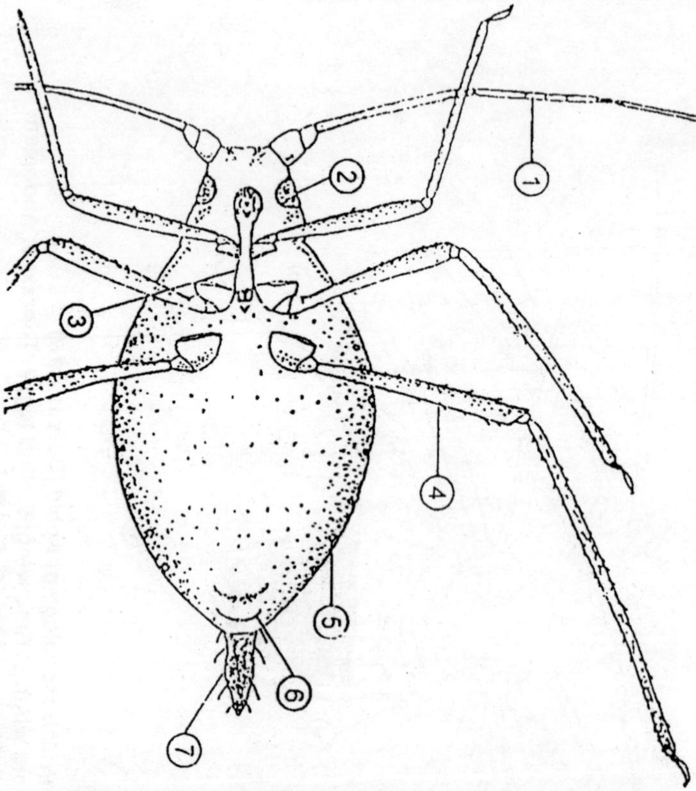

Figure 3: External Morphology of Aphid (Ventral View)
1: Antenna; 2: Eye; 3: Rostrum; 4: Hind leg; 5: Spiracle; 6: Subanal plate 7: Cauda

Eyes are present on lateral margins of the head, they may be many or three facets. When they are many faceted there may be an ocular tubercle composed of three facets only. The posterior side of eye has ommatidium may or may not be present in certain forms.

Rostrum (Figure 6) arises from the ventral surface of head. It consists of five segments, 4th and 5th segment may be fused together or separate, called ultimate rostral segment (u. r. s.). Ultimate rostral segment has variable in length and shape and it bear variable primary and secondary hairs.

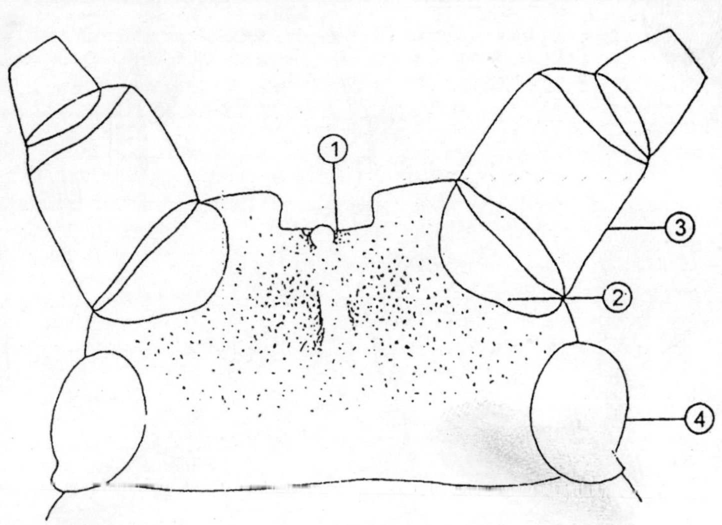

Figure 4: Head of Aphid
1: Fron; 2: Antennal socket; 3: Antenna; 4: Eye

Thorax

Thorax consists of three segments namely prothorax, mesothorax and metathorax. The prothorax or entire thorax may be fused with head or with abdomen also. Sometime mesothorax and metathorax are also fused, it may be sculptured or smooth. The prothorax bear pair of lateral tubercles, which may or may not present in mesothorax and metathorax. The mesothorax and metathoracic segments are with pair of round or oval spiracles. Mesothoracic segment has midthoracic furca.

Each thoracic segment is with one pair of legs (Figure 9), usually consists of five parts namely; coxa, trochanter, femur, tibia, tarsus and claw, some times coxa and trochanter may be smooth or sparsely hairy. The femur and tibia may be smooth or imbricate and their surface may be smooth or scabrous; hind tibiae in oviparae are usually with round pseudosenseria. Tibia is sometimes with wax pores or plates. Tarsus is usually comprised of two segment or only one segment. Second tarsal segment is provided with a pair of claws, have empodium. Claws are normally paired but may be entirely

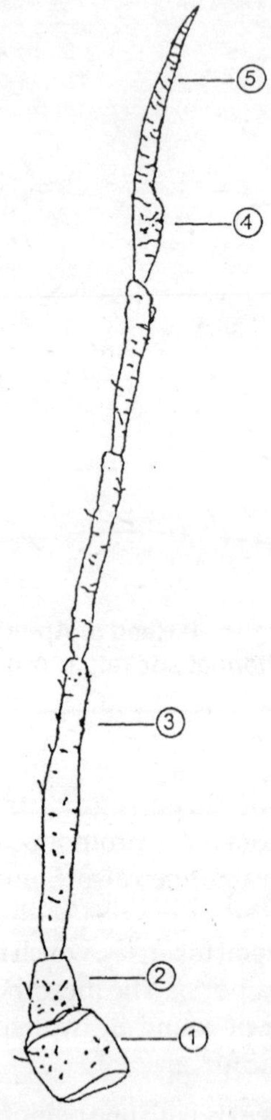

Figure 5: Antenna of Aphid
1: Scape; 2: Pedicel; 3: Flagellum; 4: Base; 5: Processus terminalis

absent or even single claw may be present. Both tarsal segments may be spinulose. Hairs on femur, tibia and tarsus vary with size, they may be sparsely or densely, thicker or thinner and fine or swollen.

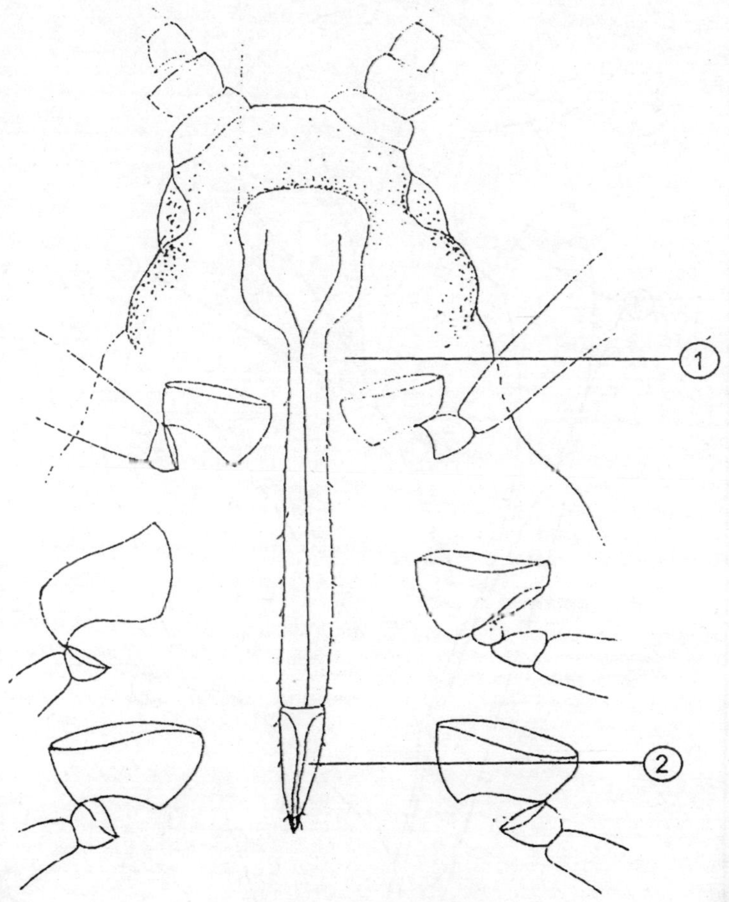

Figure 6: Rostrum of Aphid
1: Rostrum; 2: Ultimate rostral segment

Mesothoracic and metathoracic segments contain pairs of wings, namely fore and hind wings. The fore wings (Figure 9) are always longer and broader than hind wings. The fore wing has pterostigma, vein, "c" may or may not extend upto the apex of wing, radical sector (Rs) may or may not touch pterostigma; may be straight or curved. The media (M) in fore wing is simple and may be one or

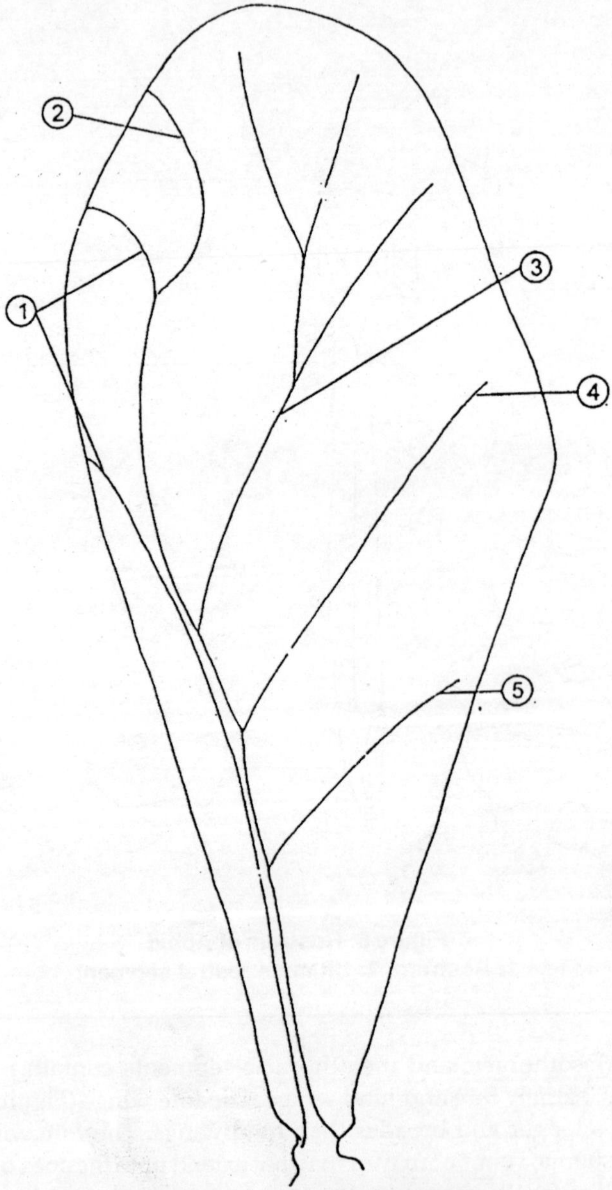

Figure 7: Fore Wing of Aphid
1: Pterostigma; 2: Radical sector (RS); 3: Media (M); 4: Cubitus (Cu); 4: Anal (A)

two branched. The cubitus (Cu) and anal vein (A) is simple, may be fused or far apart.

Hind wings (Figure 8) are very much reduced with two oblique veins *viz.* cubitus (Cu) and anal vein (A), some time both veins may be absent or fused originated from common base or apart.

Abdomen

The abdomen consists of nine segments, including last segment cauda. The first of these segments may be fused with thorax or may remain distinct. Segments may be clearly or indistinctly demarcated or may get fused. Abdominal tergites may be membranous or partly or completely sclerotic; fergile may be smooth or sculptured; dorsal hairs are very short or very long. Abdominal segments (1 to 7) are with a pair of spiracles laterally. The lateral abdominal tubercles are present on segments 1- 7. The tubercle may or may not bear hairs. Wax pores or plates may be present. Siphunculi or cornicles (Figure 10) are present on dorsolateral side of 6th abdominal segment or there are all together absent. When present they may be with various size and shapes namely; ringform, mammiform, cone shaped, cylindrical, truncate, tapering etc. Its surface may be smooth, warty and may sometimes bear some spinules on either the whole surface, may bear wax pores or plates. The opening of siphunculi may be placed right at the apex or shifted at laterally. The margin of siphunculus is smooth, imbricate or denticulate or warty. Hairs on siphunculi if present may be found all over densely or sparsely. The cauda (Figure 11) is various shaped such as semioval, oval to pentagonal, elongate or knobbed with variable size colours and with few to many hairs. The subanal plate situated ventral to cauda and may be entire or slightly to distinctly indented or bilobed and with stiff hairs. Sometimes the body terminates a tube like structure and then separate entities of cauda, subanal and subgenital plates are not distinguishable.

The superfamily Aphidoidea is characterised by,

1. Minute forms.
2. Polymorphic forms.
3. Alatae or apterous forms; alatae adults with 3 to 6 segmented antennae, first two short, rest with placoid sensoria.

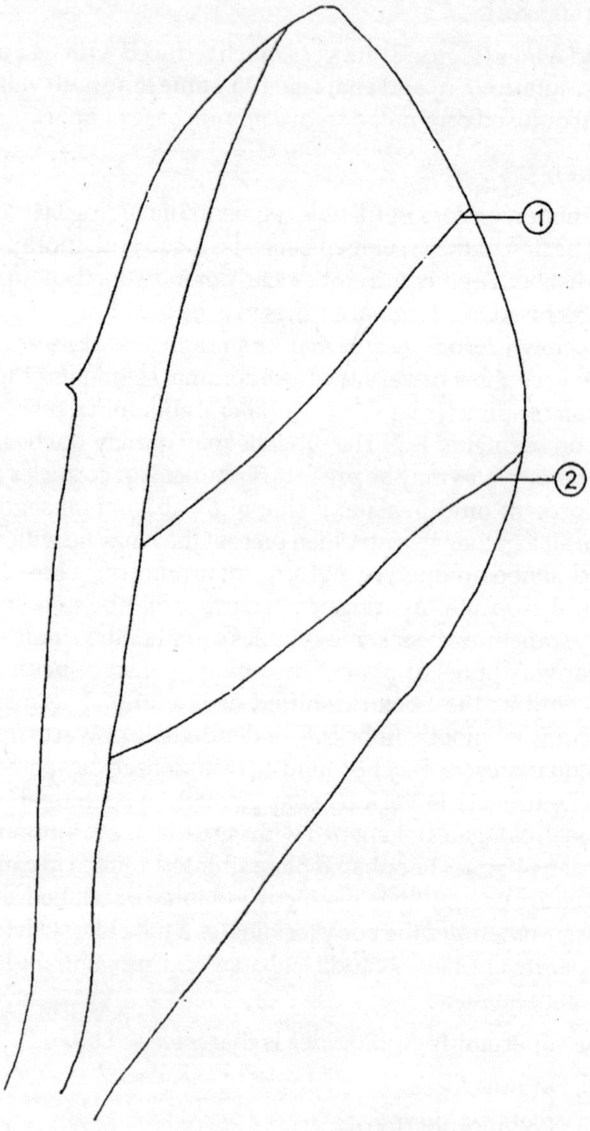

Figure 8: Hind Wing of Aphid
1: Cubitus (Cu); 2: Anal (A)

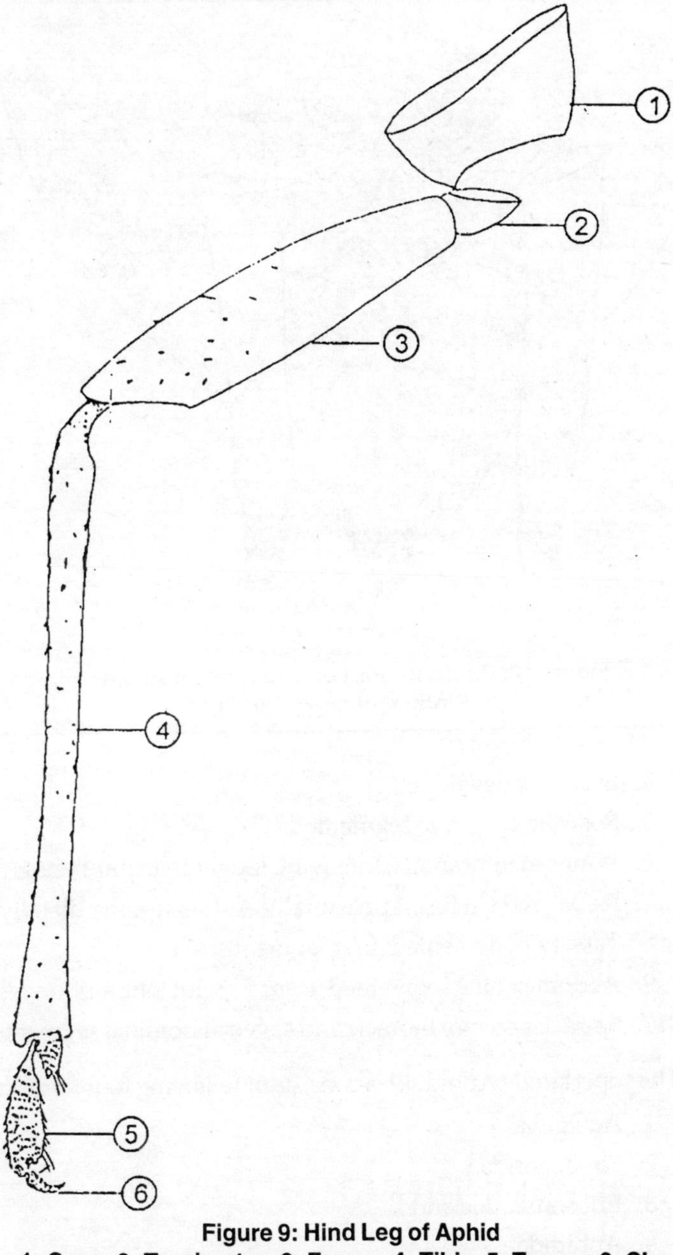

Figure 9: Hind Leg of Aphid
1: Coxa; 2: Trochantor; 3: Femur; 4: Tibia; 5: Tarsus; 6: Claw

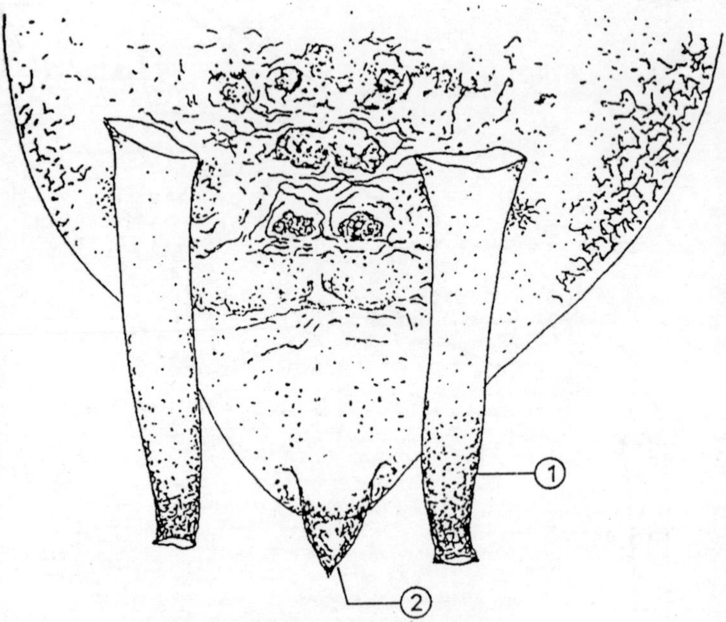

Figure 10: Posterior Part of Abdomen of Aphid
1: Siphunculus; 2: Cauda

4. Eyes well developed.
5. Rostrum long, five segmented.
6. Wings membranous, fore wing longer than hind wing.
7. Rs, M, and Cu fused proximally and separating distally.
8. Tarsi two segmented, first segment short.
9. Abdomen nine segmented, 6th or 7th with long siphunculi.
10. Spiracles on two thoracic and seven abdominal segments.

The superfamily Aphidoidea consists of following four families:

1. Adelgidae
2. Phyllozeridae
3. Eriosomatidae and
4. Aphididae.

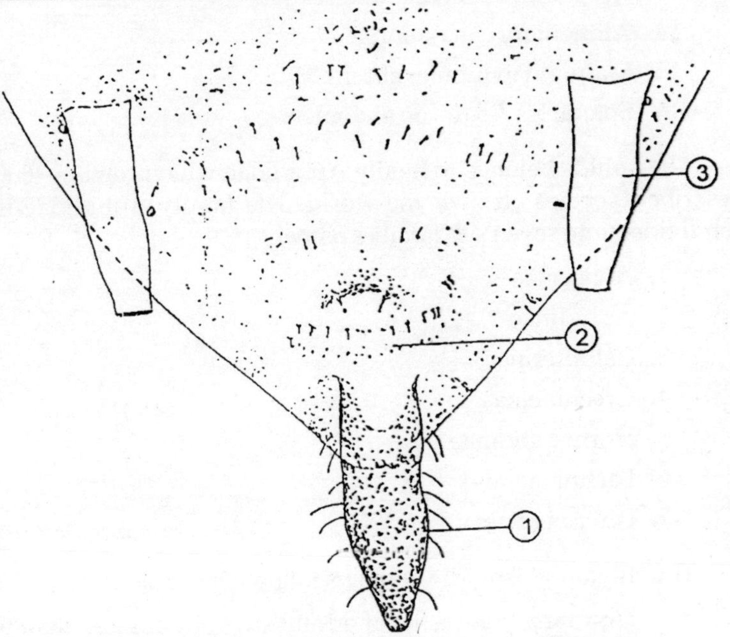

Figure 11: Posterior Part of Abdomen of Aphid
1: Cauda; 2: Subanal plate; 3: Siphunculus

The family Aphididae is characterised by following characteristics:

1. Minute or small forms.
2. Soft bodied and polymorphic.
3. Winged or wingless (Alatae and Apterous).
4. Antennae three to six segmented.
5. Sensoria on terminal segment.
6. Rostrum five segmented and long.
7. Tarsi two segmented.
8. Abdomen usually with pair of cornicles.
9. Tibia swollen in sexual females.
10. Ocelli three.
11. Compound eyes, well developed.
12. The forns often with waxy bloom or rarely stationary.

13. Reproduction aganic and sexual.
14. Alimentary canal simple.
15. Malpighian tubules absent.
16. Spicracles 7 pairs on abdomen.

The aphids belongs to family Aphididae which contain 4000 described species all over the world. The family Aphididae is subdivided into seven sub families namely:

1. Anoecinae
2. Aphidinae
3. Callipterinae
4. Greenideinae
5. Hormaphidinae
6. Lachninae and
7. Pemphiginae.

The subfamily Aphidinae shows following characters:

1. First tarsal segments of adults always with less than 9 ventral hairs and sometimes also with a pair of dorsal hairs.
2. Head may or may not have a median surface.
3. Tergite 8 and cauda usually normal, head in apterae usually distinct from pronotum, if rarely fused than eyes 3–faceted.
4. Empodial hairs always fine.
5. Head and pronotum in apterae mutually separate.
6. Processus terminalis usually longer than base of last antennal segment; if rarely shorter than even siphunculi elongate.
7. Eyes always multifaceted.
8. Siphunculi and cauda usually elongate, but when reduced p. t. longer than base of last antennal segment.
9. Body sometimes with processi, secondary rhinaria never subannular or ring like.
10. Subanal plate never indented or bilobed.

The subfamily Aphidinae is subdivided into three tribes *viz.* Aphidini, Macrosiphini and Pterocommatini.

The tribe Aphidini is characterized by:

1. Antennae and body densely pilose, apterae sometimes with secondary rhinaria.

The tribe Aphidini consists of 8 genera namely *Toxoptera* Koch, *Casimira* Eastop, *Rhopalosiphum* Koch, *Schizaphis* Borner, *Hysteroneura* Davis, *Melanaphis* Van der Goot, *Hyalopterus* Koch and *Aphis* Linnaeus.

The tribe Macrosiphini is confirmed by:

1. Lateral frontal tubercles usually pronounced.
2. Abdominal segment 1 and 2 possess spiracles, placed very close together, if apart, distance not more than 3.0 x diameter of spiracular aperture.

The tribe Macrosiphini contain 71 genera namely *Acutospiohon* (Basu), *Acyrthosiphon* (Mordvilko), *Akkaia* (Takahashi), *Amphicercidus* (Destland), *Amphorophora* (Backton), *Arthromyzus* (Borner), *Aulacorthum* (Mordvilko), *Brachycaudus* (Van der Goot), *Brachimyzus* (Basu), *Brevicoryne* (Van der Goot), *Capitophorus* (Van der Goot), *Cavariclla* (Guercio), *Chaetomyzus* (Ghosh and Raychaudhuri), *Colarodoa* (Wilson), *Cryptosiphon* (Oestlund), *Cryptosiphum* (Buckton), *Dactynotus* (Rafinsque), *Diphorodon* (Borner), *Dysaphis* (Borner), *Eomyzus* (Takahashi), *Eumyzus* (Shinji), *Hayhurstia* (Gluercio), *Hillerislambersia* (Basu), *Hyadaphis* (Kirkarldy), *Hyalomyzus* (Richards), *Hyperomyzus* (Borner), *Impatientinum* (Mordvilko), *Indaphis* (Basu), *Indomyzus* (Ghosh), *Indumasonaphis* (Raychaudhuri), *Jacksonia* (Theobald), *Liosomaphis* (Walker), *Lipaphis* (Mordvilko), *Macromyzella* (Ghosh), *Macromyzus* (Takahashi), *Macrosiphoniella* (Guercio), *Macrosiphum,* (Passirini), *Masonaphis* (Lamber), *Matsmurja* (Schumacher), *Metopolophium* (Mordvilko), *Micromyzodium* (David), *Micromyzus* (Goot), *Myzakkaia* (Basu), *Myzus* (Passirini), *Neoacyrthosiphon* (Tao), *Neohylamyzus* (Basu), *Neomegaurophis* (Ghosh), *Neomyzus* (Goot), *Oedisiphum* (Goot), *Ovatus* (Goot), *Paezoskia* (Mordvilko), *Pentalonia* (Coquerel), *Perillaphis* (Takahashi), *Pleotrichophorus* (Borner), *Pseudaphis* (Lambers), *Pseudoacythosiphon* (Ghosh), *Rhodobium* (Lambers), *Rhopalosiphonius* (Baker), *Scleromyzus* (Basu), *Semiaphis* (Goot), *Shinjia* (Takahashi),

Sinomegoura (Takahashi), *Subovatomyus* (Basu), *Taiwanomyzus* (Tao), *Tricaudatus* (Narzikulvov), *Trichosiphonaphis* (Takahashi), *Tuberoaphis* (Tao), *Tuberosephalus* (Shinji), *Vesiculaphis* (Guercio), *Walgreniella* (Lambers) and *Xenomyzus* (Aizenberg).

The tribe Pterocommatini is characterised by:

1. Antennae and body densely pilose.
2. Apterae sometimes with secondary rhinaria, usually on Salicaceae.

The tribe Pterocommatini contain only one genus *Pterocomma* (Buckton).

The subfamily Greenideinae shows following characters:

1. First tarsal segments of adults always with less than 9 ventral hairs and sometimes also with pair of dorsal hairs.
2. Head may or may not have median suture.
3. Tergite 8 and cauda shortened.
4. Head in apterae usually fused with pronotum.
5. Second tarsal segment usually dorsoapical hairs have expanded apices.
6. First tarsal segment usually with 5 or 7 ventral hairs; if number of such hairs is less, posterior abdominal segments (7–8) with long hair bearing processes.
7. Siphunculi usually long and exceptionally short truncate cones.
8. Posterior abdominal processi present.
9. In fore wing Cu and A veins placed apart basally.
10. Secondary rhinaria in alatae round and oval to transversely elongate.

The subfamily Greenideinae is subdivided into two tribes namely Cervaphidini and Greenideini.

The tribe Cervaphidini is characterised by:

1. The body of apterae with posterior abdominal processi.
2. In alatae, 'M' of forewing is once branched and hindwing with or without one or two oblique veins.
3. Siphunculi usually with very fine hair.

The tribe Cervaphidini contain 4 genera namely *Anomalosiphum* (Takahashi), *Cervaphis* (Goot), *Schoutendenia* (Rubsaamen) and *Sumatraphis* (Takahashi).

The tribe Greenideini is characterised by:

1. Eyes always multifaceted.
2. Apical tibial hairs always different from other tibial hairs.
3. Siphunculi always densely hairy

The tribe Greenideini contain 7 genera namely, *Allotrichosiphum* (Takahashi), *Brevitrichosiphon* (Raychaudhuri), *Eutrichosiphum* (Essia and Kuwana), *Greenidea* (Schoutendon), *Greenideoidea* (Goot), *Holotrichosiphum* (Raychaudhuri) and *Mollitrichosiphum* (Suenaga).

The subfamily Hormaphidinae shows following characters:

1. First tarsal segment of adults always with less than 9 ventral hairs, sometimes with pair of dorsal hairs.
2. Head may or may not have median suture.
3. Tergite 8, cauda usually shortened.
4. Head in apterae usually fused with pronotum.
5. Dorsoapical hairs on second tarsal segment usually with expanded apices.
6. First tarsal segment usually with less than 5 ventral hairs; sometimes such hairs upto 8 when hind tibiae of apterae with pseudorhinaria like structure.
7. Tibiae, tarsi densely spinulose.
8. Cauda, subanal plate is entire.
9. Siphunculi never more than a truncated cone
10. Posterior abdominal processi never present.
11. Eyes in apterae always 3- faceted.
12. Subanal plate rarely entire and more frequently slightly to distinctly bilobed.
13. Cu and A veins of fore wings placed close basally.
14. Secondary rhinaria in alatae nearly ring like or transversely oval.

The subfamily Hormaphidinae contain 20 genera namely *Aleurodaphis* (Goot), *Astegopteryx* (Karsh), *Cerataphis* (Lichtenstein),

Ceratoglyphina (Goot), *Chaitoregma* (Lambers), *Euthoracaphis* (Takahashi), *Glyphinaphis* (Goot), *Indomipponaphis* (Ghosh), *Kurisakia* (Takahashi), *Metanipponaphis* (Takahashi), *Neoceratovacuna* (Ghosh), *Neothoracaphis* (Matsmura), *Neotuberaphis* (Pal), *Nipponaphis* (Pergande), *Paraoregma* (Ghosh), *Pseudoastegopteryx* (Ghosh), *Pseudoregma* (Doncaster), *Thoracaphis* (Goot), *Teberaphis* (Takahashi) and *Ceratovacuna* (Zehntner).

GENUS *APHIS* LINNAEUS

The genus *Aphis* was erected by Linnaeus in 1758 with the help of type species *Aphis sambuci* Linnaeus. This is cosmopolitan genus. The genus shows following characteristics:

1. Head possess very weakly developed or without any lateral frontal tubercles; frons straight or slight convex, smooth.
2. Antenna usually 6 segmented (sometimes 5 segmented in apterae), shorter or longer than the body, flagellum faintly or distinctly imbricated, secondary rhinaria absent on 3rd segment of apterae, present and roundish on the alatae, sometimes secondary rhinaria is on 4th segment and 5th segment also, flagellar hairs usually short and sparse; primary rhinarium ciliated.
3. Rostrum may reach upto hindcoxa of metathoracic leg; ultimate rostral segment (u.r.s.) shorter or slight longer than hind tarsus 2 (h.t.2), u.r.s. bears secondary hairs may shorter or longer than the primary hairs.
4. Midthoracic furca without a stalk.
5. Alatae with a dark legs, F.T.C. 3, 3, 3 or 3, 3, 2.
6. Wing venation normal, veins sometimes bordered brown.
7. Dorsum of abdomen is smooth or rugose, sometimes with polygonal reticulations, in few apterae dorsal sclerotic patches and pigmented areas; dorsal hairs short with blunt spines but may quite long with acute apices.
8. Siphunculus subcylindrical, usually imbricated and dark, rarely smooth and pale.

From India, eleven species have been reported under this genus. Most of the above species are reported from West Bengal.

APHIS ACACIAE SP.NOV.

FEMALE: APTERA (Figures 12–17)

1.72 mm long, 0.88 mm broad, dark brown, oval; head 0.16 mm long, 0.36 mm broad; antenna 6 segmented, 1.20 mm long, shorter than body; thorax 0.48 mm long, 0.68 mm broad; abdomen 1.06 mm long, 0.86 mm broad, dark brown; siphunculus 0.38 mm long, 0.08 mm broad, dark brown; cauda triangular, 0.16 mm long, 0.24 mm broad, dark brown.

HEAD (Figure 12)

0.16 mm long, 0.36 mm broad, dark drown, smooth, lateral frontal tubercles weakly developed; frons with slightly convex median prominence; dorsal cephalic hairs very short, sparsely hairy with acute apices; eyes lateral with indistinct triommatidia, ocular space 0.07 mm, interocular space 0.30 mm.

Antenna (Figure 14)

1.20 mm long, pale brown, imbricated, 6 segmented, shorter than body; scape stout, 0.06 mm long, 0.05 mm broad, dark brown, sparsely hairy; pedicel short, 0.04 mm long, 0.04 mm broad, dark brown, sparsely hairy; flagellum 4 segmented, faintly imbricated, pale brown, sparsely hairy; 1st segment 0.31 mm long, 0.03 mm broad, secondary rhinaria absent; 2nd flagellar segment 0.25 mm long, 0.03 mm broad; 3rd segment 0.18 mm long, 0.02 mm broad, segment 2nd and 3rd much shorter than 4th segment; 4th segment 0.34 mm long, 0.02 mm broad, pale brown, faintly imbricated, 4th segment base 0.10 mm long, 0.02 mm broad, primary rhinaria ciliated situated at tip of base; processus terminalis (p.t.) 0.24 mm long, pale brown, imbricated, sparsely hairy, p.t. longer than base, 2.4 x base of last antennal segment; antenna shorter than body, 0.69 x body.

Flagellar Formula

1 L/W = 10.33, 2 L/W = 8.35, 4 L/W = 17.0, A = 11.88

Rostrum (Figure 12)

0.38 mm long, 0.08 mm broad, dark brown, short, sparsely hairy, extend upto hindcoxa of metathoracic leg, 5 segmented; 4th and 5th segment fused to from ultimate rostral segment (u.r.s.), u.r.s. 0.14 mm long, 0.05 mm broad, dark brown with 3 primary, 3 secondary hairs

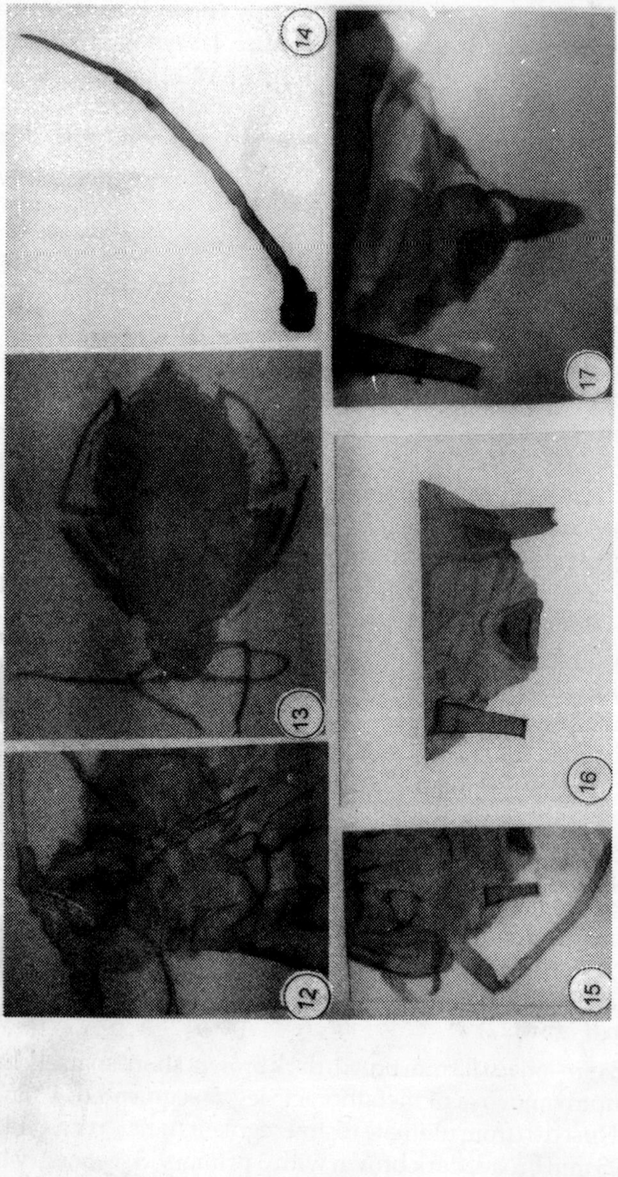

Figures 12-17: *Aphis acaciae* sp.nov. (Female Aptera)
Figure 12: Head and Rostrum; Figure 13: Whole Mount; Figure 14: Antenna; Figure 15: Hind Leg; Figure 16: Siphunuli; Figure 17: Cauda

with acute apices, secondary hairs shorter than primary; u.r.s. longer than h.t.2, 2.8 x h.t. 2.

THORAX

0.48 mm long, 0.68 mm broad, dark brown, broader than long, narrow anteriorly, broad posteriorly, sparsely hairy, short hairs with acute apices; lateral tubercles indistinct, midthoracic furca sessile; prothorax 0.14 mm long, 0.36 mm broad; mesothorax 0.18 mm long, 0.52 mm broad; metathorax 0.18 mm long, 0.66 mm broad.

Hind Leg (Figure 15)

1.15 mm long, pale brown; coxa stout, 0.10 mm long, 0.08 mm broad, dark brown, sparsely hairy, broad at base and narrow at apex; trochanter short, triangular 0.06 mm long, 0.06 mm broad, dark brown, sparsely hairy; femur 0.29 mm long, 0.08 mm broad, pale brown at base, dark brown at apex, broad posteriorly, sparsely hairy with longest hair on femora shorter than its basal diameter; tibia long, 0.52 mm long, 0.05 mm broad, pale at base, dark brown at apex, densely hairy, primary and secondary hairs with acute apices; tarsus 0.15 mm long, 0.02 mm broad, 1st tarsal segment very short, 0.02 mm long, conical, pale brown, F.T.C. 3, 3, 2; 2nd tarsal segment 0.15 mm long, 0.02 mm broad, h.t. 2 slightly curved with curved two claws, h.t. 2 with secondary hairs in addition to primary, h. t. 2 shorter than u.r.s., 0.35 x u.r.s

ABDOMEN

1.06 mm long, 0.86 mm broad, dark brown, smooth with polygonal reticulations, sclerotic patches present midorsally, narrow anteriorly, broad posteriorly, dorsal hairs very short with acuminate apices, 8 segmented; lateral abdominal tubercles on 5th and 6th segment; 1st segment 0.12 mm long, 0.55 mm broad; 2nd segment 0.14 mm long, 0.65 mm broad; 3rd segment 0.16 mm long, 0.75 mm broad; 4th segment 0.17 mm long, 0.86 mm broad; 5th segment 0.14 mm long, 0.80 mm broad; 6th segment 0.13 mm long, 0.70 mm broad; 7th segment 0.11 mm long, 0.55 mm broad; 8th segment 0.10 mm long, 0.40 mm broad with two hairs on 8th tergite, hairs short with acuminate apices; subanal plate oval, dark brown, densely spinulose.

Siphunculus (Figure 16)

0.38 mm long, 0.08 mm broad, dark brown, densely imbricated with transverse striations at base, siphunculus with indistinct apical flange at tip with densely serrated margin; siphunculus longer than cauda, 1.58 x cauda and shorter than body, 0.22 x body.

Cauda (Figure 17)

0.16 mm long, 0.24 mm broad, dark brown, elongated, densely spinulose with 6 primary, 6 secondary hairs with acute apices; cauda shorter than siphunculus, 0.38 x siphunculus.

HOST PLANTS

Khair-*Acacia catechu;* bean: *Dolichos lablab* Linn; Cow pea- *Vigna catjang.*

DISCUSSION

Aphis acaciae sp. nov. runs close to *Aphis gossypii* by having following characters.

1. Two hairs on tergite 8 become short.
2. Second tarsal segment with secondary hairs in addition to primary hairs.
3. Secondary hairs on u.r.s. shorter than primary hairs.

However, it differs from above species by having following characters.

1. Antennal segment 4th + 5th much shorter than 6th segment.
2. Proportion siphunculus to cauda.
3. Siphunculus longer than third segment of antenna.
4. Proportion antenna and siphunculus.
5. Lateral frontal tubercles on head weakly developed, frons with slight convex median prominence.
6. Flagellar formula: 1 L/W = 10.33, 2 L/W = 8.35, 4 L/W = 17.0, A = 11.88.

CONTROL MEASURES

Natural Control

Ladybird beetles: *Coccinella* sp. and Lace wings feed on the pest.

Chemical Control

Spray: 0.03 per cent phosphamidon (or) 0.03 per cent malathion (or) 0.03 per cent Azadiractin or 0.1 per cent Carbaryl.

APHIS CRACCIVORA KOCH

FEMALE: APTERA (Figures 18–23)

1.95 mm long, 1.05 mm broad, greenish black, oval, sclerotized; head 0.25 mm long, 0.44 mm broad; antenna 6 segmented, 1.28 mm long, shorter than body; thorax 0.65 mm long, 0.88 mm broad; abdomen 1.10 mm long, 1.05 mm broad; siphunculus 0.41 mm long, 0.04 mm broad, dark greenish black; cauda 0.19 mm long, 0.08 mm broad, dark greenish black.

HEAD (Figure 18)

0.25 mm long, 0.44 mm broad, dark greenish black with separate sclerite, lateral frontal tubercles absent; frons with convex median prominence, sparsely hairy with acute apices; eyes lateral with distinct triommatidia, ocular space 0.10 mm, interocular space 0.36 mm.

Antenna (Figure 20)

1.28 mm long, pale greenish black, imbricated, with sparsely hairy, 6-segmented; scape stout, 0.6 mm long, 0.08 mm broad, dark greenish black, without hairs; pedicel short, straight, 0.05 mm long, 0.05 mm broad, dark greenish black, without hairs; flagellum 4 segmented; 1st segment 0.35 mm long, 0.05 mm broad, pale greenish black, distinctly imbricated without secondary rhinaria; 2nd segment 0.23 mm long, 0.03 mm broad; 3rd segment 0.22 mm long, 0.03 mm broad; 4th segment 0.39 mm long, 0.02 mm broad, pale greenish black, densely imbricated, 4th segment base 0.11 mm long, dark greenish black, distinctly imbricated, ciliated round primary rhinarium at tip of base; processus terminalis (p.t.) 0.26 mm long, densely imbricated, sparsely hairy with acute apices, p.t. longer than base, 2.26 x base of last antennal segment; antenna shorter than body, 0.66 x body.

Flagellar Formula

1 L/W= 7.00, 2 L/W = 7.6, 4 L/W =19.5, A=11.36.

Rostrum (Figure 18)

0.38 mm long, 0.07 mm broad, greenish black, sparsely hairy, 5 segmented; 4th and 5th segment fused to form ultimate rostral segment (u.r.s.); u.r.s. 0.13 mm long, 0.04 mm broad, u.r.s. with secondary as long as primary hairs with acute apices; u.r.s. shorter than h.t.2, 0.68 x h.t.2.

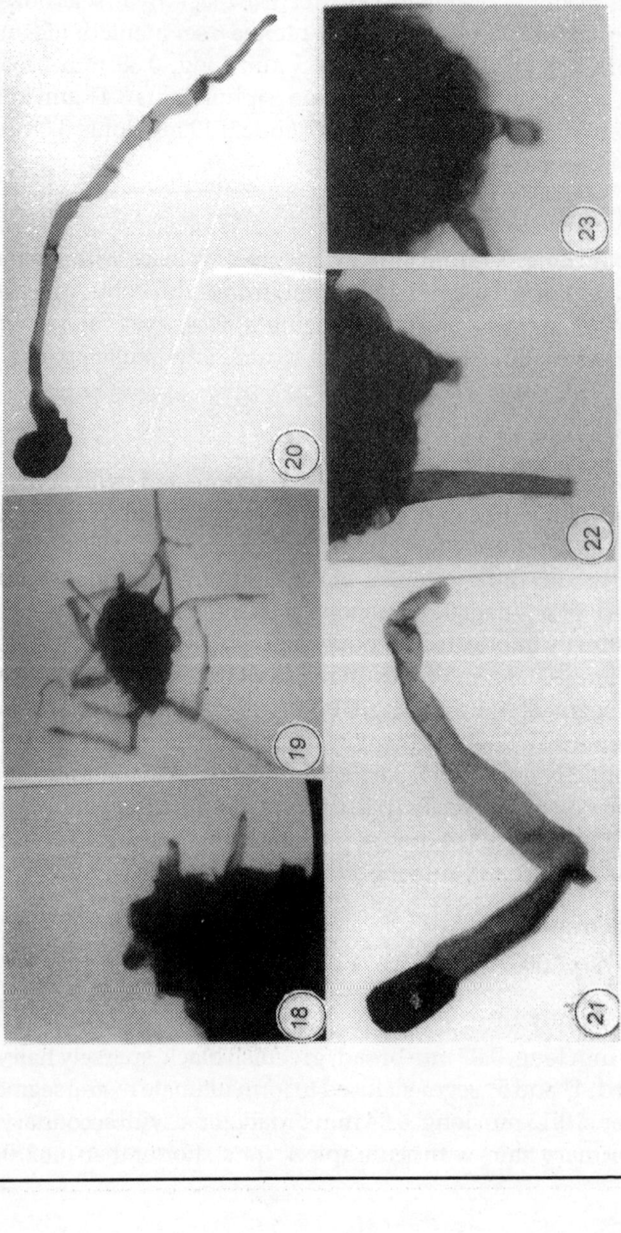

Figures 18-23: *Aphis craccivora* (Female Aptera)
Figure 18: Rostrum; Figure 19: Whole Mount; Figure 20: Antenna; Figure 21: Hind Leg; Figure 22: Siphunculi; Figure 23: Cauda

THORAX

0.62 mm long, 0.89 mm broad, dark greenish black, sclerotic tranverse bands present, without hairs; prothorax 0.19 mm long, 0.55 mm broad; mesothorax 0.24 mm long, 0.73 mm broad; metathorax 0.18 mm long, 0.89 mm broad.

Hind Leg (Figure 21)

1.92 mm long, dark greenish black; coxa stout, 0.09 mm long, 0.10 mm broad, sparsely hairy, broad at base, narrow at apex; trochanter short, straight, 0.07 mm long, 0.06 mm broad, sparsely hairy; femur 0.60 mm long, 0.06 mm broad, pale greenish black at base, dark at apex, densely imbricated, broad posteriorly, medium hairs with acute apices, longest hairs on femora shorter than basal diameter of femora; tibia long, straight; 1.05 mm long, 0.05 mm broad, pale greenish black at base, dark at apex with dense secondary and primary hairs with acute apices; tarsus 0.19 mm long, 0.02 mm broad, 1st tarsal segment 0.02 mm long, pale greenish black, conical, F. T. C. with 3, 3, 3 fine tarsal hairs; 2nd tarsal segment (h. t. 2) 0.19 mm long, slightly curved with curved two claws with normal imbrications; h.t.2 longer than u.r.s., 1.46 x u.r.s.

ABDOMEN

1.10 mm long, 1.05 mm broad, dark greenish black, dorsum completely sclerotic, pale area around siphunculi with distinctly polygonal reticulations, sparsely hairy with acute apices, 8 segmented; 1st segment 0.10 mm long, 0.85 mm broad; 2nd segment 0.11 mm long, 0.92 mm broad; 3rd segment 0.13 mm long, 0.95 mm broad; 4th segment 0.18 mm long, 1.05 mm broad; 5th segment 0.14 mm long, 0.89 mm broad; 6th segment 0.15 mm long, 0.58 mm broad; 7th segment 0.13 mm long, 0.49 mm broad; 8th segment 0.10 mm long and 0.15 mm broad; subanal plate oval, dark greenish black with medium sized hairs.

Siphunculus (Figure 22)

0.41 mm long, 0.04 mm broad, dark greenish black, without hairs, faintly imbricated with distinct polygonal reticulations with distinct apical flange; siphunculus longer than cauda, 2.73 x cauda and shorter than body, 0.25 x body.

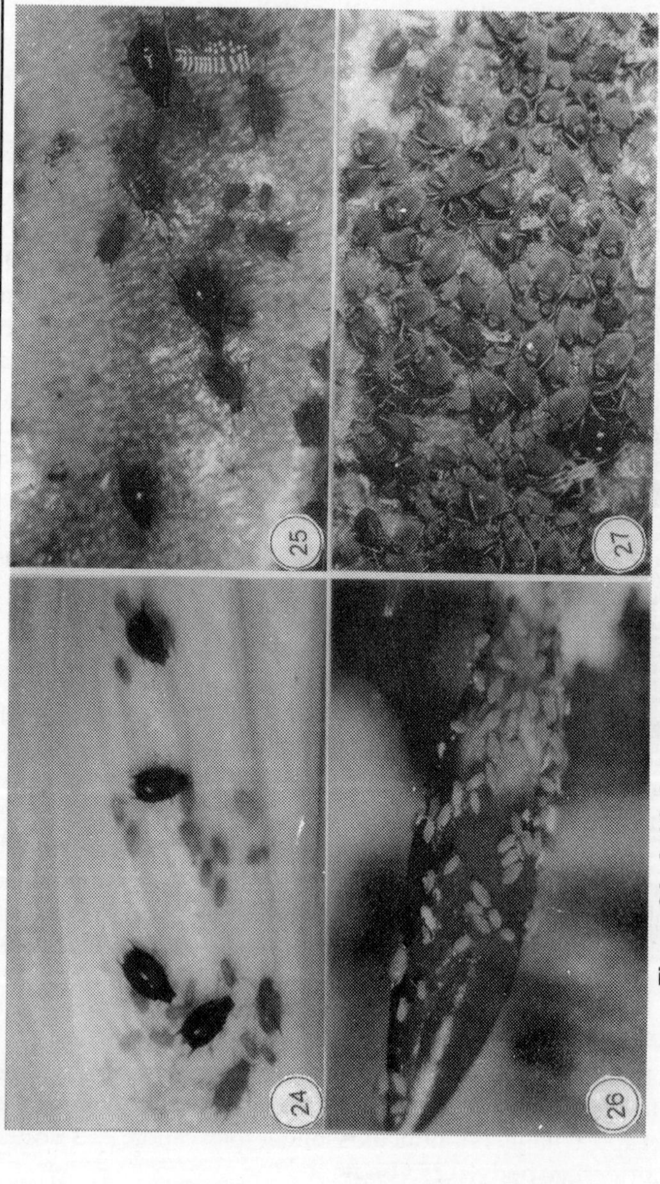

Figure 24: *Aphis jaswandi* sp.nov.; Figure 25: *Aphis acaciae* sp.nov.; Figure 26: *Aphis kaneri* sp.nov.; Figure 27: *Aphis craccivora*

Cauda (Figure 23)

0.19 mm long, 0.08 mm broad, elongated, dark greenish black, tapered at tip, 5 medium sized hairs with acute apices; cauda shorter than siphunculus, 0.36 x siphunculus.

HOST PLANTS

Cow pea: *Vigna catjang* Walp.; Mango–*Mangifera indica*; Hirda–*Terminalia chebula*; Jamun–*S. cumunis*; Ground nut–*Arachis hypogea*, Sunflower–*Helianthus annus*; Pigeon pea–*Cajanus cajan*.

CONTROL MEASURES

Natural Control

Ladybird beetles: *Menochilus* sp., *Coccinella* sp. and *Ishiodon javana* (Syphridae: Diptera).

Chemical Control

Spray: 0.08 per cent Thiometon (or) 0.03 per cent Metasystox (or) 0.04 per cent Monocrotophos; Dusting: 5 per cent Malathion or 3 per cent Morphothion.

APHIS GOSSYPII GLOVER

FEMALE: APTERA (Figures 28–34)

1.56 mm long, 0.68 mm broad, pale greenish brown, elongated; head 0.18 mm long, 0.35 mm broad, pale greenish brown, smooth; antenna 6 segmented, 0.81mm long, shorter than body; thorax 0.49 mm long, 0.56 mm broad; abdomen 0.82 mm long, 0.69 mm broad, pale greenish brown; siphunculus 0.22 mm long, 0.05 mm broad, dark green brown; cauda 0.09 mm long, 0.06 mm broad, dark greenish brown.

HEAD (Figure 28)

0.18 mm long, 0.35 mm broad, pale greenish brown, smooth; lateral frontal tubercles weakly developed; frons with slight convex median prominence; eyes lateral with distinct triommatidia, ocular space 0.09 mm, interocular space 0.28 mm.

Antenna (Figure 31)

0.81 mm long, pale green, smooth, 6 segmented; scape stout, 0.05 mm long, 0.07 mm broad, dark greenish brown, smooth; pedicel

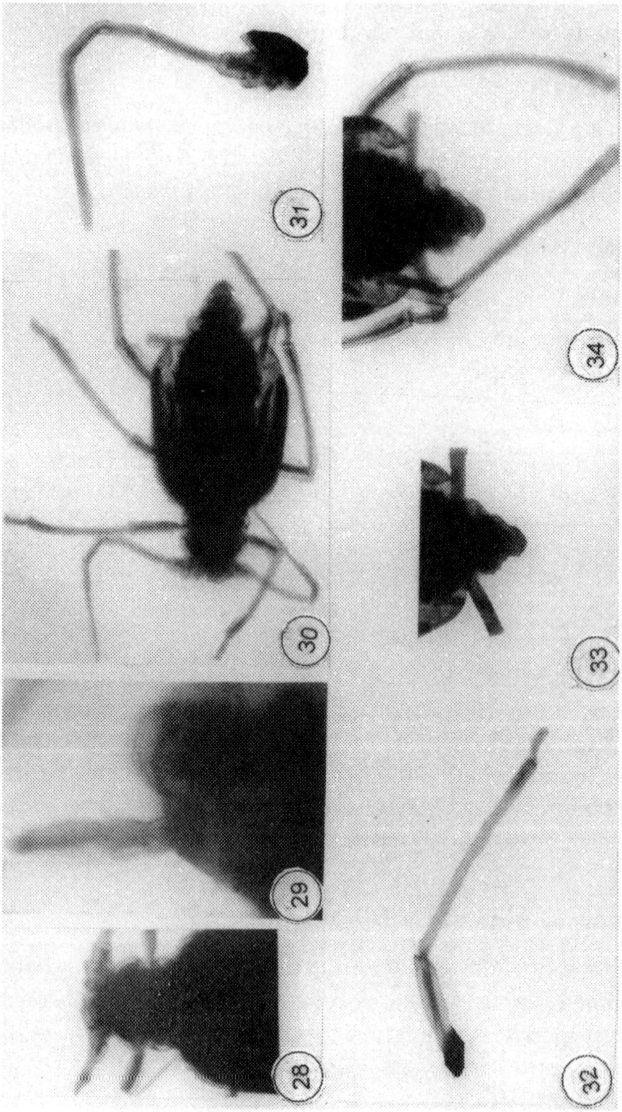

Figure 28-34: *Aphis gossypii* sp.nov. (Female Aptera)
Figure 28: Head; Figure 29: Rostrum; Figure 30: Whole Mount; Figure 31: Antenna;
Figure 32: Hind Leg; Figure 33: Siphunculi; Figure 34: Cauda

short, 0.04 mm long, 0.06 mm broad, dark greenish brown, smooth; flagellum 4 segmented, slight distinctly imbricated, sparsely hairy with acute apices; 1st segment 0.17 mm long, 0.03 mm broad, pale greenish brown, smooth, without secondary rhinaria; 2nd segment 0.14 mm long, 0.02 mm broad; 3rd segment 0.13 mm long, 0.02 mm broad; 4th segment 0.34 mm long, 0.03 mm broad, dark greenish brown, imbricated, 4th segment base 0.08 mm long, 0.03 mm broad with ciliated primary rhinarium at tip of base; processus terminalis (p.t.) 0.26 mm long, 0.02 mm broad, p.t. longer than base, 3.3 x base of last antennal segment; antenna shorter than body, 0.16 x body.

Flagellar Formula

1 L/W = 5.6, 2 L/W = 7.00, 4 L/W = 17.0, A= 9.86.

Rostrum (Figure 29)

0.35 mm long, 0.06 mm broad, pale green, sparsely hairy, extend upto hindcoxa of metathoracic segment, 5 segmented; 4th and 5th segments fused to from ultimate rostral segment (u.r.s.), u.r.s. 0.12 mm long, 0.02 mm broad, dark green, sparsely hairy, secondary hairs on u.r.s. shorter than primary hairs with acute apices; u.r.s. longer than h.t. 2, 1.09 x h.t. 2.

THORAX

0.49 mm long, 0.56 mm broad, pale greenish brown, smooth, without hairs, wider than the long; prothorax 0.15 mm long, 0.32 mm broad, smooth, without hairs, prothorax separate from head and mesothorax; mesothorax 0.22 mm long, 0.53 mm broad, smooth, without hairs; metathorax 0.13 mm long, 0.56 mm broad, smooth, without hairs; mesothorax continuous with metathorax.

Hind Leg (Figure 32)

0.98 mm long, pale greenish brown; coxa stout, 0.06 mm long, 0.07 mm broad, dark greenish brown, sparsely hairy, broad at base, narrow at tip; trochanter short, 0.04 mm long, 0.06 mm broad, dark green, sparsely hairy; femur long, 0.28 mm long, 0.02 mm broad, slightly curved, pale at base, dark at apex sparsely hairy, longest hair on femora shorter than basal diameter of femora; tibia long, straight, 0.48 mm long, 0.02 mm broad, dense primary, secondary hairs with acute apices, tibia pale at base, dark at apex; tarsus 0.11 mm long, 0.02 mm broad, pale greenish brown, 1st tarsal segment short, 0.02 mm long, conical, pale greenish brown, F. T. C. with 3, 2,

3 fine hairs; 2nd tarsal segment (h.t.2) 0.11 mm long, dark green, slightly curved with two claws with secondary hairs in addition to primary hairs; secondary hairs at ventral side, h.t. 2 slight shorter than u.r.s, 0.91 x u.r.s.

ABDOMEN

0.82 mm long, 0.69 mm broad, pale greenish brown, smooth, rarely sclerotic patches at middle of dorsal, sparsely hairy with acuminate apices, 8 segmented; spiracles at lateral side of 1- 7 segment, 1st segment 0.07 mm long, 0.62 mm broad; 2nd segment 0.08 mm long, 0.64 mm broad; 3rd segment 0.09 mm long, 0.65 mm broad; 4th segment 0.11 mm long, 0.69 mm broad; 5th segment 0.10 mm long, 0.59 mm broad; 6th segment 0.12 mm long, 0.28 mm broad; 7th segment 0.07 mm long, 0.14 mm broad, 8th segment 0.09 mm long, 0.09 mm broad, 2 hairs on tergite 8 short blunt to acuminate apices and shorter than 3rd segment; 4th and 5th segment always slightly shorter than segment 6th.

Siphunculus (Figure 33)

0.22 mm long, 0.05 mm broad subcylindrical, dark greenish brown, faintly imbricated, smooth, without hairs; siphunculus longer than 3rd segment of antenna; siphunculus longer than cauda, 2.44 x cauda and shorter than body, 0.14 x body.

Cauda (Figure 34)

0.09 mm long, 0.06 mm broad, dark greenish brown, conical, 4 medium sized slight curved hairs at tip; cauda shorter than siphunculus, 0.40 x siphunculus.

HOST PLANTS

Cotton: *Gossypium arboreum;* Mango–*Mangifera indica; Eucalyptus sp.;* Ain–*Terminalia tomentosa;* Jamun–*S. cumunis;* Pigeon pea–*Cajanus cajan;* Tomato–*Lycopersicon esculeatum;* Rose–*Rose multiforma.*

CONTROL MEASURES

Natural Control

Ladybird beetle: *Menochilus* sp. and *Coccinella* sp. and lacewings, hover fly, etc. feed on pest and suppress population.

Chemical Control

Spray: 0.02 per cent phosphamidon (or) 0.02 per cent methyl demeton (or) 0.02 per cent Rogor (or) 0.02 per cent DDVP (or) 0.03 per cent Azadirachtin.

Dusting: 5 per cent malathion.

GENUS *MELANAPHIS* VAN DER GOOT

The genus *Melanaphis* was erected by Van der Goot in 1917 with the help of type species *Aphis bambussae* Fullaway. The cosmopolitan genus is distributed in India, Africa, Australia, China, Egypt, Hawaii, Iraq, Japan, Java, Sri Lanka and Taiwan. The genus is visualized by following features.

1. Head with weakly developed lateral frontal tubercles but shows "rhopalosiphine" projection without median frontal prominence, dorsal longitudinal striae in apterae smooth or indistinctly to distinctly stippled dorsally.
2. Antenna 5 or 6 segmented in apterae, 6 segmented in alatae; about 0.40–1.10 x and 0.65–0.93 x body of apterae and alatae respectively; weakly or strongly imbricated, p.t. about 1.42–4.50 x and 2.70–4.20 x base of last antennal segment of apterae and alatae, secondary rhinaria absent inapterae, but inalatae circular, secondary rhinaria restricted on 3rd segment or may be on 4th and 5th segment, antennal hairs apterae like dorsal cephalic hairs but alatae acuminate shorter to longer than basal diameter of 3rd segment.
3. Rostrum extend upto midcoxa of mesothoracic legs, 5 segmented, 4th and 5th segment fused to form u.r.s. with only two secondary hairs, u.r.s. shorter than or may be up to 1.90 x h.t. 2.
4. Legs pale, femur smooth dorsally but spinulose striae ventrally, hairs may short or long, F.T.C. 3, 3, 2 but sometimes 3, 3, 3.
5. Wing venation normal, wing veins may or may not bordered brown, radial sector strongly or weakly curved.
6. Abdominal tergites pale or dorsal pigmentation, smooth or scabrous, dorsal abdominal hairs may short with blunt

apices, lateral abdominal tubercles developed on 2-6 segment.

7. Siphunculus weakly or strongly imbricated, either wider than long or longer than wide with or without constructions just before apex.
8. Cauda usually dark, elongate with globules to slightly acute apex and median construction, bear 4- 21 hairs, when siphunculus shorter than cauda number of caudal hairs more.
9. Subgenital plates with 2-11 hairs on its anterior half and there may be short or long.

From India, seven species have been described under this genus.

MELANAPHIS VULGARI SP.NOV.

FEMALE: APTERA (Figures 35–39)

1.96 mm long, 0.95 mm broad, pale brown; head 0.15 mm long, 0.40 mm broad; antenna 6 segmented, 0.86 mm long, shorter than body; thorax 0.60 mm long, 0.90 mm broad; abdomen 1.20 mm long, 0.95 mm broad; siphunculus short, 0.10 mm long, 0.04 mm broad, slight dark brown; cauda short, 0.08 mm long, 0.04 mm broad, slight dark brown.

HEAD (Figure 36)

0.15 mm long, 0.40 mm broad, pale brown, faintly scabrous, lateral frontal tubercles weakly developed; frons without median frontal prominence, dorsal cephalic hairs very short, sparsely with bluntish apices; eyes lateral with distinct triommatidia, ocular space 0.08 mm long, interocular space 0.32 mm.

Antenna (Figure 37)

0.86 mm long, pale brown, sparsely hairy; 6 segmented, shorter than body; scape stout, 0.05 mm long, 0.06 mm broad, pale brown, sparsely hairy with blunt apices; pedicel short, 0.04 mm long, 0.04 mm broad, pale brown, sparsely hairy with blunt apices; flagellum 4 segmented, sparsely hairy with bluntish apices, 1st segment 0.17 mm long, 0.04 mm broad, without secondary rhinaria; 2nd segment 0.14 mm long, 0.03 mm broad; 3rd segment 0.11 mm long, 0.02 mm broad; 4th segment 0.36 mm long, 0.02 mm broad, pale brown, 4th

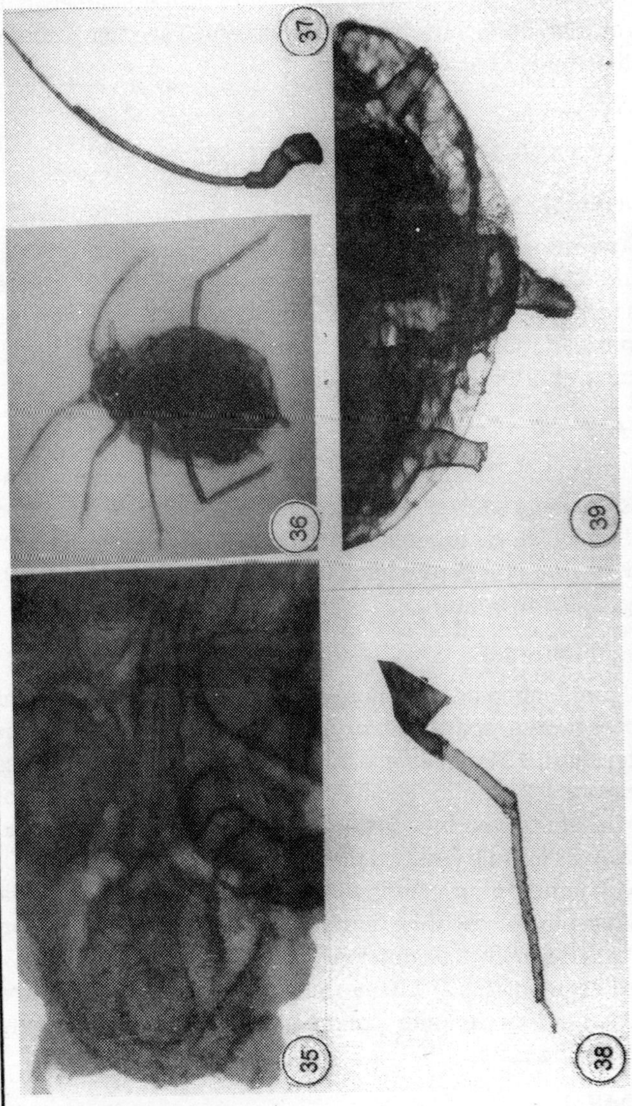

Figures 35-39: *Melanaphis vulgari* sp.nov. (Female Aptera)
Figure 35: Head and Rostrum; Figure 36: Whole Mount; Figure 37: Antenna; Figure 38: Hind Leg; Figure 39: Siphunculi and Cauda

segment base short, 0.09 mm long, 0.02 mm broad, pale brown, primary rhinarium ciliated, round at tip of base; processus terminalis (p.t.) 0.27 mm long, pale brown, strongly imbricated, sparsely hairy; p.t. longer than base, 3.0 x base of last antennal segment; antenna shorter than body, 0.44 x body.

Flagellar Formula

1 L/W = 4.25, 2 L/W= 3.5, 4 L/W = 18.00, A = 8.5.

Rostrum (Figure 35)

0.30 mm long, 0.04 mm broad, pale brown, sparsely hairy, extend upto midcoxa of mesothoracic leg; 5 segmented; 4th and 5th segment fused to form ultimate rostral segment (u.r.s.), u.r.s. 0.08 mm long, 0.04 mm broad, dark brown with 2 secondary hairs with pointed apices, u.r.s. shorter than h.t.2, 0.80 x h. t. 2.

THORAX

0.60 mm long, 0.90 mm broad, pale brown, pigmented, faintly scabrous, sparsely hairy with acute apices; mesothorax, metathorax with indistinct lateral tubercles; prothorax 0.14 mm long, 0.45 mm broad; mesothorax 0.25 mm long, 0.74 mm broad; metathorax 0.23 mm long, 0.90 mm broad.

Hind Leg (Figure 38)

1.22 mm long, pale brown; coxa stout, 0.08 mm long, 0.10 mm broad, pale brown, sparsely hairy, broad at base, narrow at apex; trochanter short, 0.04 mm long, 0.06 mm broad, pale brown, sparsely hairy; femora 0.36 mm long, 0.06 mm broad, pale brown, smooth, spinulose, slight curved, broad posteriorly, longest hair on midfemora shorter than its middle width; tibia long, narrow, 0.60 mm long, 0.04 mm broad, faintly imbricated with dense primary and secondary hairs with acute apices, tibia densely hairy at apex; tarsus 0.10 mm long, 1st tarsal segment 0.02 mm long, pale brown, conical, F.T.C. 3, 3, 3; 2nd tarsal segment (h.t. 2) 0.10 mm long, slightly curved with curved two claws, hind tarsus with 2 short secondary hairs; h.t.2 longer than u.r.s., 1.25 x u.r.s.

ABDOMEN

1.20 mm long, 0.95 mm broad, pale brown, faintly scabrous, lateral abdominal tubercles on 2-6 segments, dorsal hairs usually short with bluntish apices, anterior 4 segments not more than one marginal hair, longest hair on anterior, abdominal tergite shorter than b.d. of 3rd segment, 8 segmented; 1st segment 0.13 mm long, 0.90 mm broad; 2nd segment 0.16 mm long, 0.92 mm broad; 3rd segment 0.15 mm long, 0.94 mm broad; 4th segment 0.17 mm long, 0.95 mm broad; 5th segment 0.15 mm long, 0.85 mm broad; 6th segment 0.16 mm long, 0.60 mm broad; 7th segment 0.08 mm long, 0.30 mm broad; 8th segment 0.08 mm long, 0.16 mm broad; subanal plate pale brown, short hairs with acute apices; subgenital plate slight dark brown, semi-oval with 7 short hairs at anterior part with acute apices.

Siphunculus (Figure 39)

0.10 mm long, 0.04 mm broad, short, slight dark brown, weakly imbricated with transverse striatations; siphunculus longer than its basal diameter, 2.5 x middle of siphunculus; siphunculus longer than cauda, 1.25 x cauda and shorter than body, 0. 51 x body.

Cauda (Figure 39)

0.08 mm long, 0.04 mm broad, slight dark brown, elongated with slight acute apex, bears 7 short hairs with acute apices; cauda shorter than siphunculus, 0.80 x siphunculus.

HOST PLANTS

Jowar: *Sorghum vulgure* Pers.

DISCUSSION

This species runs close to *Melanaphis sacchari* Zehntner by

1. Longest hair on anterior abdominal tergite shorter than b.d. of 3rd segment of antenna.
2. Siphunculi longer than its basal width.
3. Longest hair on midfemora shorter than its middle width.

However, it differs from above species by having following character:

1. Siphunculi longer than cauda.
2. Marginal hairs on anterior abdominal segment only one.
3. Lateral frontal tubercles on head weakly developed, frons without median frontal prominence.
4. Processus terminalis about thrice the length of base.

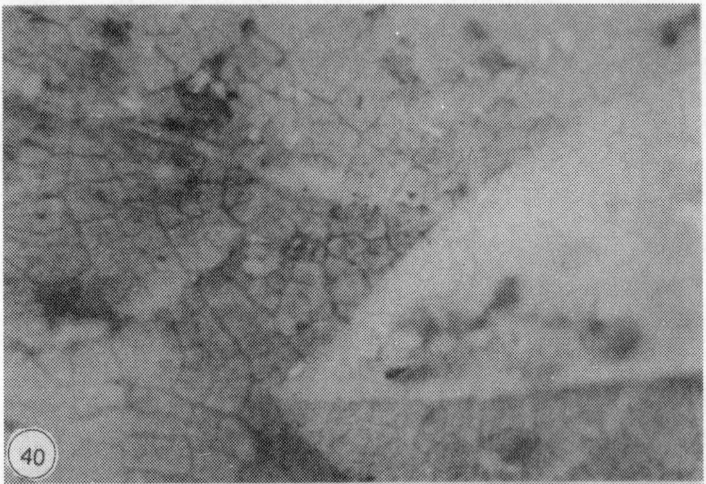

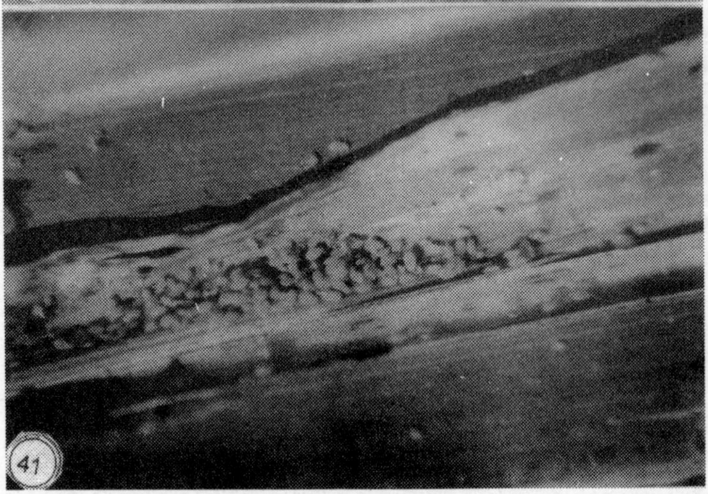

Figure 40: Colony of *Aphis gossypii*;
Figure 41: Colony of *Melanaphis vulgari* sp.nov.

5. Cauda with 7 short hairs, not more than 9 as found in *M. sacchari*.
6. Proportion antenna to siphunculus.
7. Proportion siphunculi to cauda.
8. Flagellar formula: 1 L/W = 4.25, 2L/W= 3.5, 4 L/W = 18.00, A = 8.5.

CONTROL MEASURES

Natural Control

Ladybird beetles: *Menochilus* sp., Lace wing and Syriphid larvae feed on the pest.

Chemical Control

Spray the crop with any of the following insecticide: 0.03 per cent Azadirachtin; 0.03 per cent malathion; 0.01 per cent carbaryl; 0.05 per cent DDVP.

Dusting: 5 per cent malathion 20 kg/ha

GENUS *TOXOPTERA* KOCH

The genus *Toxoptera* was erected by Koch in 1857 with the help of type species *Aphis aurantii* Boyer de Fonscolombe which is cosmopolitan in distribution. The genus shows following characters:

1. Head smooth or fine granulated on dorsum, without lateral frontal tubercles.
2. Antenna 6 segmented, shorter or slight longer than body, about 0.53–1.08 x body, flagellum distinctly imbricated from base to apex, apterae lack secondary rhinaria while alatae secondary rhinaria on 3rd segment or may be 4th and 5th segment, p.t. about 2.05–6.37 x base of last antennal segment, flagellar hairs short to long with acute, acuminate or blunt apices.
3. Rostrum extends upto hindcoxa of metathoracic leg, u.r.s. normal, nearly as long as to 1.60 x h.t. 2, with 2 long secondary hairs.
4. Midthoracic furca with a broad base.
5. Leg smooth, femur smooth with some scaly imbrications, forefemur paler than mid and hindfemur, tibia medially

pale but coloured basally and apically with scaly imbrications; hind tibiae with row of peg like stradulatory spines, F.T.C. 3, 3, 3 or 3, 3, 2.

6. Fore wings with media once or twice branched and hind wings with oblique veins.
7. In apterae, abdominal dorsum with polygonal reticulations with large post siphuncular sclerite and scattered muscle plate like structure and some dark patches marginally; in alatae pale with segmently arranged spinopleural patches, with large post siphuncular patch, transverse bars on 7 and 8 tergite, dorsal hairs with acute apices, venter shows rows of spinulose striae which are lateroventral side of 5th, 6th and 7th segment.
8. Siphunculus brown to dark brown, weakly or strongly imbricated, subcylindrical to cylindrical with indistinct to distinct apical flange, about 0.04–0.18 x body.
9. Cauda elongated, constricted at the middle with 10–46 long hairs.

From India, three species have been described under this genus.

TOXOPTERA AMBI SP. NOV.

FEMALE: APTERA (Figures 42–47)

1.60 mm long, 0.75 mm broad, dark brown (blackish); head 0.16 mm long, 0.32 mm broad; antenna 6 segmented, 1.10 mm long, shorter than body; thorax 0.65 mm long, 0.62 mm broad; abdomen 0.82 mm long, 0.75 mm broad, dark brown; siphunculus 0.27 mm long, 0.04 mm broad, dark brown; cauda 0.15 mm long, 0.06 mm broad, dark brown.

HEAD (Figure 42)

0.16 mm long, 0.32 mm broad, dark brown, scabrous, rough, sparsely hairy with acute apices; lateral frontal tubercles weakly developed; frons with weakly developed median prominence; eyes lateral with distinct triommatidia, ocular space 0.10 mm, interocular space 0.26 mm.

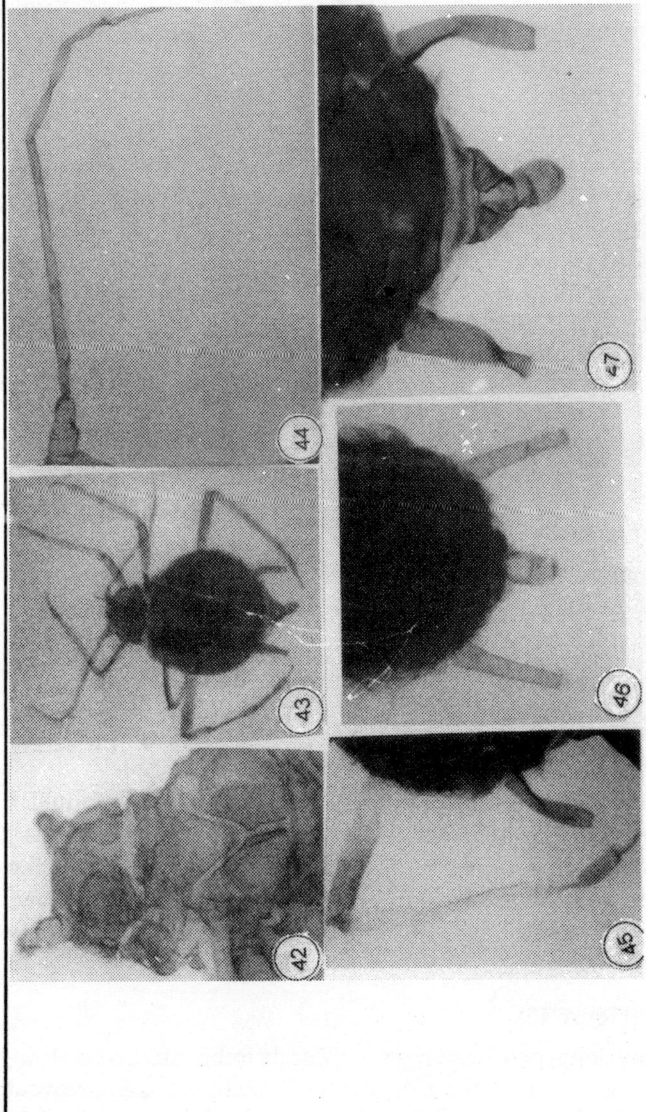

Figures 42-47: ***Toxoptera ambi*** **sp.nov. (Female Aptera)**
Figure 42: Head and Rostrum; Figure 43: Whole Mount; Figure 44: Antenna; Figure 45: Hind Leg; Figure 46: Siphunculi; Figure 47: Cauda

Antenna (Figure 44)

1.10 mm long, pale brown, densely imbricated with serrated margin, 6 segmented, shorter than body; scape stout 0.07 mm long, 0.06 mm broad, dark brown, sparsely hairy with acute apices; pedicel short, straight 0.05 mm long, 0.05 mm broad, dark brown, sparsely hairy with acute apices; flagellum 4 segmented, pale brown, densely imbricated; 1st segment 0.28 mm long, 0.04 mm broad, secondary rhinaria absent, sparsely hairy with longest hair on 1st segment, 0.25 x its basal diameter; 2nd segment 0.22 mm long, 0.04 mm broad; 3rd segment 0.20 mm long, 0.03 mm broad; 4th segment 0.38 mm long, 0.02 mm broad, pale brown, distinctly imbricated, 4th segment base short, 0.12 mm long, 0.02 mm broad, dark brown, densely imbricated with serrated margin, sparsely hairy with acute apices, primary rhinarium ciliated, oval at tip of base; processus terminalis (p.t.) 0.26 mm long, densely imbricated with serrated margin, sparsely hairy with acute apices, p.t. longer than base, 2.1 x base of last antennal segment; antenna shorter than body, 0.68 mm x body.

Flagellar Formula

1 L/W = 7.00, 2 L/W = 5.5, 4 L/W = 19, A= 10.5.

Rostrum (Figure 42)

0.37 mm long, 0.04 mm broad, dark brown, extend upto hindcoxa of metathoracic leg, sparely hairy, 5 segmented; 4th and 5th segment fused to from ultimate rostral segment (u.r.s.), u.r.s. 0.16 mm long, 0.02 mm broad, dark brown with 2 secondary hairs on u.r.s. with acute at apices; u.r.s longer than h.t.2, 1.60 x h.t. 2.

THORAX

0.65 mm long, 0.62 mm broad, dark brown with blackish tinges, rough, scabrous, sparsely hairy, midthoracic furca with a broad base; prothorax 0.15 mm long, 0.39 mm broad, lateral side with horn like outgrowth, prothorax free from head; mesothorax 0.30 mm long, 0.62 mm broad; metathorax 0.19 mm long, 0.55 mm broad; mesothorax and metathorax fused.

Hind Leg (Figure 45)

1.08 mm long, pale brown, smooth, scaly imbricated; coxa stout, 0.08 mm long, 0.08 mm broad, dark brown, broad at base, narrow at tip, as broad as long, sparsely hairy; trochanter short, straight, 0.04

mm long, 0.06 mm broad, dark brown with sparsely hairy; femur long, 0.40 mm long, 0.06 mm broad, slight curved, smooth with few scaly imbrications, pale brown at base, dark brown at apex; tibia 0.48 mm long, 0.04 mm broad, pale brown at medially but dark brown at base and apex with few scaly imbrications with dense primary and secondary hairs with acute apices; tarsus 0.12 mm long, 0.02 mm broad, 1st tarsal segment, 0.02 mm long, pale brown, conical, F.T.C. with 3,3,3; 2nd tarsal segment 0.10 mm long, 0.02 mm broad, slightly curved with curved two claws, h.t. 2 with 3 short hairs; h.t.2 shorter than u.r.s., 0.65 x u.r.s.

ABDOMEN

0.82 mm long, 0.75 mm broad, dark brown with polygonal reticulations, sclerite at postsiphuncular region scattered, muscleplate like structure with few dark patches marginally and sparsely hairy with acute apices, 8 segmented; lateral abdominal tubercles weakly developed on 1-7 segment; 1st segment 0.02 mm long, 0.06 mm broad; 2nd segment 0.12mm long, 0.69 mm broad; 3rd segment 0.13 mm long, 0. 72 mm broad; 4th segment 0.14 mm long, 0.75 mm broad; 5th segment 0.10 mm long, 0.70 mm broad; 6th segment 0.10 mm long, 0.55 mm broad; 7th segment 0.08 mm long 0.40 mm broad; 8th segment 0.06 mm long, 0.25 mm broad; subanal plate normal; subgenital plate with 25 hairs become long and short with acute apices.

Siphunculus (Figures 46 and 47)

0.27 mm long, 0.04 mm broad, dark brown, cylindrical, weakly imbricated with indistinct apical flange at tip; siphunculus longer than cauda, 1.80 x cauda; siphunculus shorter than body, 0.16 x body.

Cauda (Figure 47)

0.15 mm long, 0.06 mm broad, elongated, dark brown, slightly constricted at middle, blunt, round at apex; 12 secondary hairs with acute apices; cauda shorter than siphunculus, 0.55 x siphunculus.

HOST PLANTS

Mango–*Mangifera indica* Linn.; Neem–*Azardirachta indica;* Jamun–*S. cumunis.*

Figures 48-51: *Toxoptera ambi* sp.nov.
Figure 48: Infected Mango Twig; Figure 49: Infected Mango Leaf; Figure 50: Colony on Mango Leaf; Figure 51: Colony (Enlarged) Feeding on Mango Leaf

DISCUSSION

This species runs close to *Toxoptera aurantii* Boyer de Fonscolombe by

1. Siphunculi shorter than body, 0.10–0.17 x body.
2. Siphunculi less prominently imbricated.

However, it differs from the above species by following characters:

1. Siphunculi longer than cauda, 1.80 x cauda.
2. Proportion siphunculus to antenna.
3. Lateral frontal tubercles on head weakly developed, frons with weakly developed median frontal prominence.
4. Processus terminalis shorter, about half the length of *T. aurantii*.
5. Ultimate rostral segment acutely pointed.
6. Flagellar formula: 1 L/W = 7.00, 2 L/W = 5.5, 4 L/W = 19, A= 10.5.

CONTROL MEASURES

Natural Control

Ladybird beetles: *Coccinella* sp. and *Brumus* and lacewings feed on the pest.

Chemical Control

Spray: 0.05 per cent DDVP (or) 0.03 per cent phosphamidon (or) 0.05 per cent Azadirachtin.

Dusting: 5 per cent malathion.

GENUS *RHOPALOSIPHUM* KOCH

The genus *Rhopalosiphum* was erected by Koch in 1854 with the help of type species *Aphis nymphaceae* Linnaeus. It is cosmopolitan in distribution. This genus shows following features.

1. Head smooth with feebly developed lateral frontal tubercles shows rhopalosiphine projections.

2. Antenna 5–6 segmented, about 0.31–0.81 x body, p.t. 1.70 -6.80 x base of last segment, secondary rhinaria absent in apterae.
3. Rostrum extends upto hindcoxa of metathoracic leg, u.r.s. about 0.71–1.53 x h.t.2 with 2–4 secondary hairs.
4. Midthoracic furca in apterae usually sessile.
5. Legs brown to dark brown, coxa and femur partly imbricated, other segments usually smooth, but tibiae easily imbricated, F.T.C. 3, 3, 3 or 3, 3, 2.
6. Wing venation normal, sometimes only media (M) is once branched.
7. Abdomen dorsum of apterae pale and pleural intersegmental pigmentation in alatae; segment 7 and 8 shows well developed polygonal reticulation on dorsum of apterae, each polygon shows group of spinulose, lateral abdominal tubercles on 1- 7 segments.
8. Siphunculus dark, cylindrical to slight swollen with well developed flange about 0.05–0.25 x body.
9. Cauda short, elongated, constricted at the middle, about 0.35–0.93 x siphunculus with 10 to 46 long hairs.
10. Subgenital plate pale brown to dark brown transversely oval, with many hairs.

From India five species have been described under this genus.

RHOPALOSIPHUM CASHIVI SP. NOV.

FEMALE: APTERA (Figures 52–56)

2.10 mm long, 0.95 mm broad, dark chocolate; head 0.19 mm long, 0.40 mm broad; antenna 6 segmented, 1.05 mm long, shorter than body; thorax 0.64 mm long, 0.69 mm broad; abdomen 1.25 mm long, 0.95 mm broad; siphunculus, 0.12 mm long, 0.04 mm broad, dark chocolate; cauda 0.10 mm long, 0.12 mm broad, dark chocolate.

HEAD (Figure 52)

0.19 mm long, 0.40 mm broad, dark chocolate, slight scabrous, oval with shiny wax pores pale on dorsum, sparsely hairy with fine apices; frons with weak developed median prominence, weakly

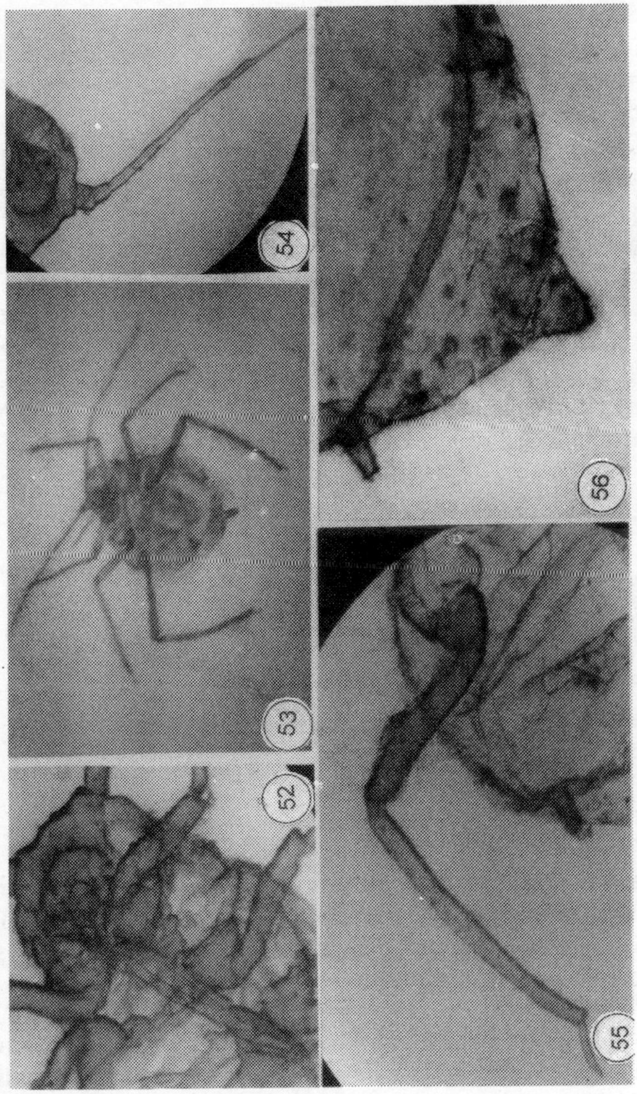

Figures 52-56: *Rhopalosiphum cashivi* sp.nov. (Female Aptera)
Figure 52: Head and Rostum; Figure 53: Whole Mount; Figure 54: Antenna; Figure 55: Hind Leg; Figure 56: Siphunculi and Cauda

developed lateral frontal tubercles diverging; eyes lateral with distinct triommatidia, ocular space 0.09 mm, interocular space 0.33 mm.

Antenna (Figure 54)

1.05 mm long, pale chocolate, 6 segmented, shorter than body; scape stout, 0.08 mm long, 0.08 mm broad, as long as wide, dark chocolate, sparely hairy, hairs short; pedicel short, straight, 0.06 mm long, 0.06 mm broad, dark chocolate sparsely hairy, hairs short; flagellum 4 segmented, pale chocolate, imbricated; 1st segment 0.23 mm long, 0.04 mm broad, without secondary rhinaria; 2nd segment 0.16 mm long, 0.04 mm broad; 3rd segment 0.16 mm long, 0.04 mm broad; 4th segment 0.34 mm long, 0.02 mm broad, 4th segment base 0.09 mm long, 0.02 mm broad, dark chocolate, primary rhinaria ciliated, round at tip of base; processus terminalis (p.t.) 0.28 mm long, slight pale, imbricated with serrated margin, sparsely hairy with acute apices; p.t. longer than base, 3.1 x base of last antennal segment; antenna shorter than body, 0.45 x body.

Flagellar Formula

1L/W = 5.75, 2 L/W = 4.00, 4 L/W= 19.00, A= 9.58.

Rostrum (Figure 52)

0.41 mm long, 0.06 mm broad, pale chocolate, extend upto hind coxa of metathoracic leg, 5 segmented; 4th and 5th segment fused to from ultimate rostral segment (u.r.s), u.r.s. 0.12 mm long, 0.04 mm broad, dark chocolate, pointed at apex with 2 short secondary hairs with acute apices.

THORAX

0.64 mm long, 0.69 mm broad, dark chocolate, scabrous, sparsely hairy, hairs very short with acute apices, midthoracic furca sessile; prothorax 0.19 mm long, 0.39 mm broad; mesothorax 0.25 mm long, 0.65 mm broad; metathorax 0.20 mm long, 0.69 mm broad.

Hind Leg (Figure 55)

Hind leg 1.10 mm long, pale chocolate, imbricated; coxa short, 0.08 mm long, 0.08 mm broad, dark chocolate, sparsely hairy with acute apices, broad at base, narrow at tip; trochanter short, 0.04 mm long, 0.06 mm broad, dark chocolate, sparsely hairy with acute apices; femur 0.26 mm long, 0.06 mm broad, slight curved, imbricated, pale at base, dark at apex; tibia long, straight 0.52 mm long, 0.04 mm broad, imbricated with dense primary and secondary hairs with acute apices; tarsus 0.10 mm long, 0.03 mm broad; 1st tarsal segment short, 0.02 mm long, pale chocolate, conical, F.T.C. with 3,3,3; 2nd tarsal segment 0.10 mm long, slightly curved with curved two claws, h.t.2 with long, fine hairs, h.t. 2 shorter than u.r.s., 0.83 x u.r.s.

ABDOMEN

1.25 mm long, 0.95 mm broad, dark chocolate, broad anteriorly, narrow posteriorly with polygonal reticulation, each polygon encloses a group of spinules, hairs short, long with fine apices, 8 segmented; lateral abdominal tubercles indistinct on 1-7 segments; 1st segment 0.14 mm long, 0.80 mm broad; 2nd segment 0.14 mm long, 0.84 mm broad; 3rd segment 0.13 mm long, 0.86 mm broad; 4th segment 0.12 mm long, 0.95 mm broad; 5th segment 0.13 mm long, 0.85 mm broad; 6th segment 0.14 mm long, 0.60 mm broad; 7th segment 0.16 mm long, 0.46 mm broad; 8th segment 0.16 mm long 0.24 mm broad; subanal plate entire, pale with many long hairs; subgenital plate dark chocolate with many long hairs with acute apices.

Siphunculus (Figure 56)

0.12 mm long, 0.04 mm broad, cylindrical, smooth, imbricated, slight broad at base with distinct apical flange; siphunculus longer than cauda, 1.2 x cauda and shorter than body, 0.57 x body.

Cauda (Figure 56)

0.10 mm long, 0.12 mm broad, slight round, dark chocolate, with 2 long, 7 short hairs with acute apices; cauda shorter than siphunculus, 0.83 x siphunculus.

HOST PLANTS

Cashew nut: *Anacardium occidentale* Linn.

DISCUSSION

This species runs close to *Rhopalosiphum maidis* Fitch by

1. Dorsal abdominal hairs with fine apices.
2. Antennae 6 segmented.

However, it differs from above species by having following characters.

1. Processus terminalis 3.1 x base of last antennal segment.

2. Body 17.50 x siphunculi.
3. Lateral frontal tubercles on head weakly developed, frons with weakly developed median frontal prominence.
4. Dorsal abdominal hairs not blunt with fine apices.
5. Siphunculi much longer than *R. maidis,* thrice the length of *R. maidis*.
6. Cauda much longer than *R. maidis*, thrice the length of *R. maidis*.
7. Flagellar formula: 1L/W = 5.75, 2 L/W = 4.00, 4 L/W= 19.00, A= 9.58.

CONTROL MEASURES

Natural Control

Ladybird beetles: *Menochilus* sp. feed on the pest population.

Chemical Control

Spray: 0.03 per cent phosphamidon (or) 0.1 per cent Carbaryl; 0.3 per cent malathion; 0.03 per cent Azadirachtin.

RHOPALOSIPHUM KOLHAPURENSIS SP. NOV.

FEMALE: APTERA (Figures 57–62)

1.78 mm long, 0.89 mm broad, dark brown; head 0.21 mm long, 0.36 mm broad; antenna 6 segmented, 0.89 mm long, shorter than body; thorax 0.62 mm long, 0.75 mm broad; abdomen 0.98 mm long, 0.89 mm broad, dark brown; siphunculus 0.22 mm long, 0.04 mm broad, dark brown; cauda 0.14 mm long, 0.22 mm broad, dark brown.

HEAD (Figure 57)

0.21 mm long, 0.36 mm broad, dark brown, smooth, sparsely hairy; wax pores oval, pale brown and various sized; lateral frontal tubercles weakly developed; frons weakly developed median prominence; eyes lateral with indistinct triommatidia, ocular space 0.08 mm, interocular space 0.31 mm.

Antenna (Figure 59)

0.89 mm long, pale brown, sparsely hairy, 6 segmented, shorter than body; scape stout, 0.05 mm long, 0.06 mm broad, dark brown,

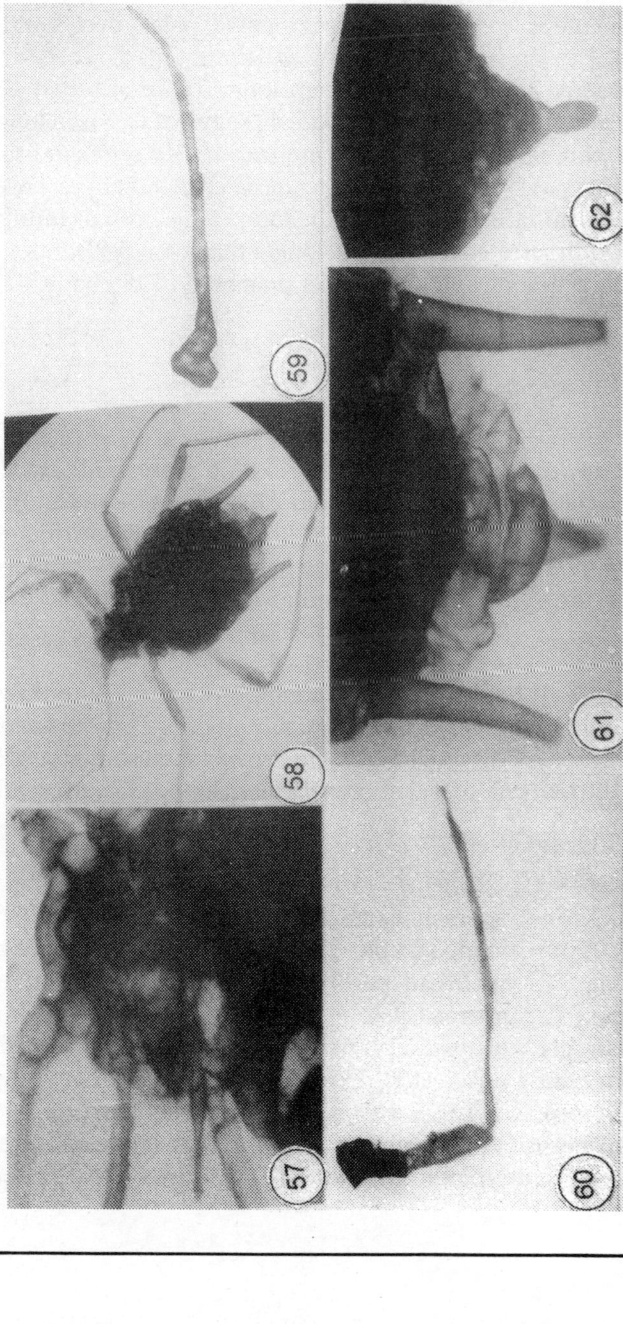

Figures 57-62: ***Rhopalosiphum kolhapurensis*** **sp.nov. (Female Aptera)**
Figure 57: Head and Rostrum; Figure 58: Whole Mount; Figure 59: Antenna;
Figure 60: Hind Leg; Figure 61: Siphunculi; Figure 62: Cauda

sparsely hairy with acute apices, broad at base, tapered at apex; pedicel short, 0.04 mm long, 0.04 mm broad, dark brown, sparsely hairy with acute apices; flagellum 4 segmented, pale brown, faintly imbricated; 1st segment 0.18 mm long, 0.04 mm broad, secondary rhinaria absent; 2nd segment 0.15 mm long, 0.03 mm broad; 3rd segment 0.14 mm long, 0.02 mm broad; 4th segment 0.38 mm long, 0.02 mm broad, pale brown, faintly imbricated, 4th segment base 0.08 mm long, pale brown, primary rhinaria ciliated at tip of base; processus terminalis (p.t.) 0.30 mm long, pale brown, faintly imbricated with serrated margin, p.t. longer than base, 3.60 x base of last antennal segment, antenna shorter than body, 0.18 x body.

Flagellar Formula

1 L/W = 4.5, 2 L/W = 5.0, 4 L/W = 19, A= 9.5.

Rostrum (Figure 57)

0.30 mm long, 0.05 mm broad, extend upto midcoxa of mesothoracic leg, 5 segmented; 4th and 5th segment fused to form ultimate rostral segment (u.r.s.), u.r.s. 0.10 mm long, 0.04 mm broad, dark brown, pointed at tip with 6 long primary, 6 short secondary hairs with acute apices; u.r.s. shorter than h.t.2, 0.83 x h.t.2.

THORAX

0.62 mm long, 0.75 mm broad, dark brown, narrow anteriorly, broad posteriorly, sparsely hairy with acute apices, lateral tubercles distinct; prothorax 0.12 mm long, 0.40 mm broad; mesothorax 0.25 mm long, 0.60 mm broad; metathorax 0.25 mm long, 0.75 mm broad.

Hind Leg (Figure 60)

1.36 mm long, pale brown; coxa stout, 0.10 mm long, 0.08 mm broad, dark brown, sparsely hairy; trochanter short, narrow, 0.05 mm long, 0.05 mm broad, dark brown, sparsely hairy; femora long, 0.32 mm long, 0.04 mm broad, pale brown, partly imbricated; tibia 0.56 mm long, 0.03 mm broad, pale brown, rarely imbricated, densely hairy with slight blunt apices; tarsus 0.16 mm long, 0.02 mm broad, pale brown, faintly imbricated; 1st tarsal segment, 0.04 mm long, pale brown, conical, F.T.C. with 3,3,3; 2nd tarsal segment 0.12 mm long, slightly curved with curved two claws, h.t. 2 with dorsolateral long hairs with acute apices; hind tarsus (h.t.2) longer than u.r.s, 1.2 x h.t.2.

ABDOMEN

0.98 mm long, 0.89 mm broad, dark brown, broad anteriorly, tapered posteriorly; wax pores oval, shiny, pale brown, various sized, present throughout abdomen; dorsal hairs medium, sparsely hairy with fine apices; 8 segmented; lateral abdominal tubercles indistinct on 1 -7 segments; 1st segment 0.15 mm long, 0.68 mm broad; 2nd segment 0.12 mm long, 0.84 mm broad; 3rd segment 0.14 mm long, 0.89 mm broad; 4th segment 0.14 mm long, 0.85 mm broad; 5th segment 0.12 mm long, 0.70 mm broad; 6th segment 0.10 mm long, 0.55 broad; 7th segment 0.09 mm long, 0.30 mm broad; 8th segment 0.07 mm long, 0.20 mm broad; subanal plate pale brown with many long hairs; subgenital plate, oval, dark brown, many long hairs with acute apices.

Siphunculus (Figure 61)

0.22 mm long, 0.04 mm broad, dark brown, cylindrical, densely reticulated with serrated margin, slight broad at base, tapered at apex with distinct apical flange at tip; siphunculus longer than cauda, 1.57 x cauda and shorter than body, 0.12 x body.

Cauda (Figure 62)

0.14 mm long, 0.22 mm broad, dark brown, elongated, semioval, moderately spinulose with 3 long hairs with bifurcated apices; cauda shorter than siphunculus, 0.63 mm x siphunculus.

HOST PLANTS

Grass: family–Graminae; Pigeon pea–*Cajanus cajan*.

DISCUSSION

This species runs close to *Rhopalosiphum padi* Linnaeus by

1. Cauda 0.63 x siphunculi.
2. Cauda with never more than 3 hairs.
3. Processus terminalis never less than 3.0 x base of last antennal segment.

However, it differs from above species by having following characters.

1. Dorsal abdominal hairs with fine apices.
2. Processus terminalis 3.60 x base of 6th segment.

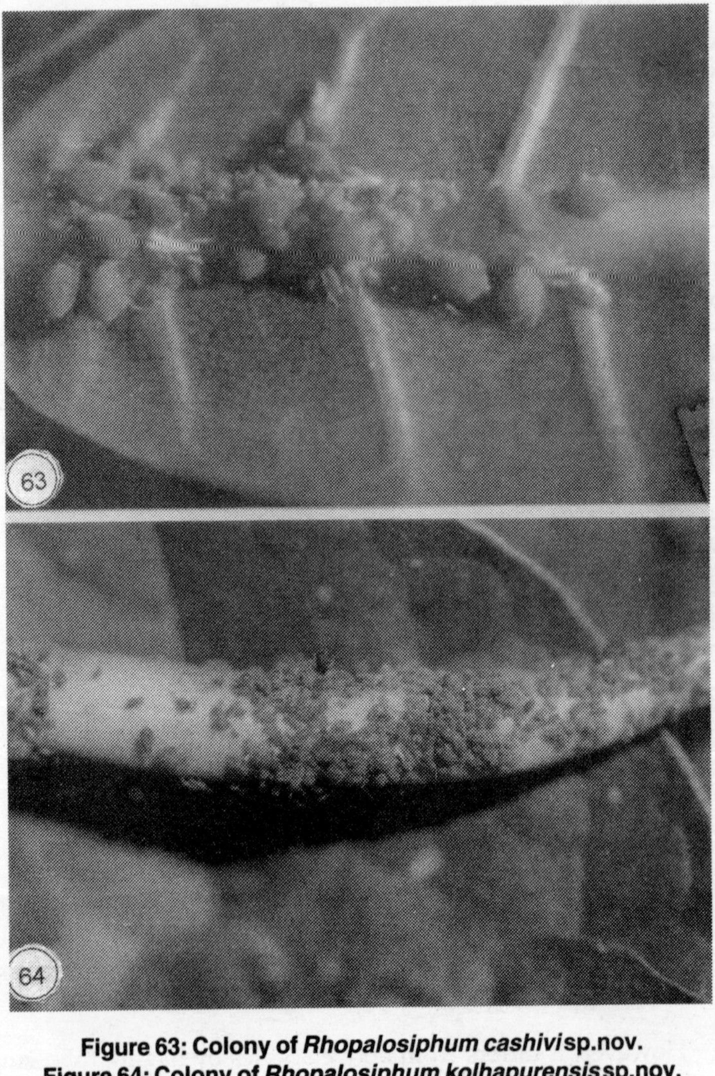

Figure 63: Colony of *Rhopalosiphum cashivi* sp.nov.
Figure 64: Colony of *Rhopalosiphum kolhapurensis* sp.nov.

3. Ultimate rostral segment longer than h.t. 2.
4. Proportion of siphunculus and antenna.
5. Proportion of siphunculus and cauda.
6. Lateral frontal tubercles on head weakly developed, frons with weakly developed median frontal prominence.
7. Flagellar formula: 1 L/W = 4.5, 2 L/W = 5.0, 4 L/W = 19, A= 9.5.

CONTROL MEASURES

Natural Control

Ladybird beetles: *Menochilus* sp., *Coccinella* sp., lacewing, suppress the pest population by feeding on aphids.

Chemical Control

Spray: 0.03 per cent phosphamidon (or) 0.03 per cent DDVP; 0.03 per cent Azadirachtin

GENUS *ACYRTHOSIPHON* MORDVILKO

The genus *Acyrthosiphon* was erected by Mordvilko in 1914. He used *Aphis pisi* Kaltenbach as type species. The genus is visualized by following characteristic features.

1. Head in apterae and alatae usually smooth, sometimes very few indistinct spinules ventrally near base of antennal sockets, well developed lateral frontal tubercles, frontal sinus in apterae 'V' shaped, dorsal cephalic hairs short with blunt apices.
2. Antenna 6 segmented, slightly shorter to fairly longer than body, imbricated, in apterae basal part of 3rd segment smooth with few secondary rhinaria, in alatae secondary rhinaria upto 25, arranged in more or less in rows, p.t. long, about 3.0–5.0 x base of last antennal segment, flagellar hairs short with blunt apices.
3. Ultimate rostral segment of rostrum shorter than h.t.2.
4. Midthoracic furca sessile.
5. Femur nearly smooth, F.T.C. 3,3,3.
6. Wing venation normal.

7. In apterae abdominal dorsum pale, smooth, in alatae only with faint spinulose marginal sclerites, dorsal hairs like dorsal cephalic hairs nearly, 0.50 x basal diameter of 3rd segment.
8. Siphunculus cylindrical with a wide base, distinct apical flange, imbricated, sometimes with a few interconnecting preapical striae about 0.20–0.25 x body.
9. Cauda elongated with pointed to blunt apices, about 0.50–0.65 x siphunculi, hairs nearly base with acute apices.
10. Subgenital plate with short blunt hairs on the hind margin which are hardly, 0.50 x basal diameter of 3rd segment.

Under the genus *Acyrthosiphon* Mordvilko 4 subgenera, namely *Acyrthosiphon*, *Microlophium* Mordvilko with *Aphis urticae* Schrank as the type, *Amphorophora* Buckton with *Amphorophora ampulata* Buckton as the type and *Metapolophium* Mordvilko with *Aphis diharda* Walker as type have been described. From the genus *Acyrthosiphon* only one species has been reported from India.

ACYRTHOSIPHON CALATROPI SP. NOV.

FEMALE: APTERA (Figures 65–70)

1.69 mm long, 0.85 mm broad, dark yellow; head 0.21 mm long, 0.39 mm broad; antenna 6 segmented, 1.25 mm long shorter than body; thorax 0.60 mm long, 0.72 mm broad; abdomen 0.90 mm long, 0.85 mm broad, dark yellow; siphunculus 0.29 mm long, 0.08 mm broad, dark yellow; cauda 0.15 mm long, 0.16 mm broad, dark yellow.

HEAD (Figure 65)

0.21 mm long, 0.39 mm broad, smooth, dark yellow, broader than long; lateral frontal tubercles well developed, diverging; frontal sinus slightly raised, U shaped; few indistinct spinulose at base of antennal socket; dorsal cephalic hairs short, sparsely hairy with incrassate apices; eyes lateral with indistinct triommatidia, ocular space 0.08 mm, interocular space 0.30 mm.

Antenna (Figure 67)

1.25 mm long, pale yellow, 6 segmented, shorter than body; scape stout, 0.08 mm long, 0.08 mm broad, dark yellow, sparsely hairy; pedicel short, straight, 0.06 mm long, 0.06 mm broad, dark

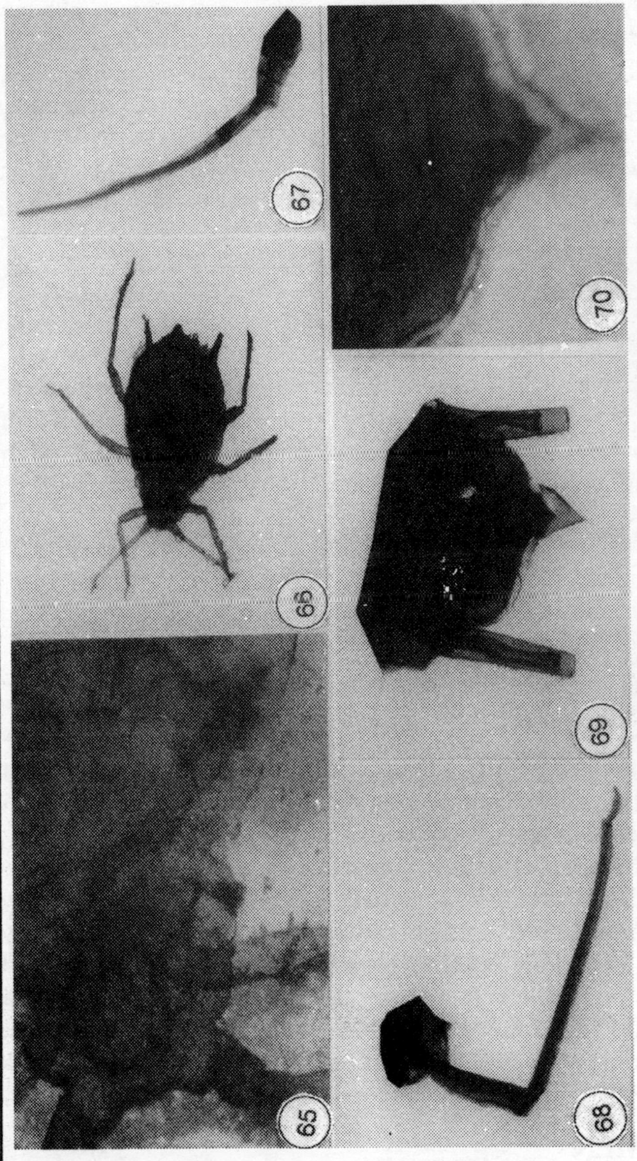

Figures 65-70: *Acyrthosiphon calatropi* sp.nov. (Female Aptera)
Figure 65: Head and Rostrum; Figure 66: Whole Mount; Figure 67: Antenna; Figure 68: Hind Leg; Figure 69: Siphunculi; Figure 70: Cauda

yellow, sparsely hairy; flagellum 4 segmented, faintly imbricated, pale yellow; 1st segment 0.28 mm long, 0.04 mm broad, with 3 secondary rhinaria at base; 2nd segment 0.28 mm long, 0.04 mm broad; 3rd segment 0.20 mm long, 0.03 mm broad; 4th segment 0.38 mm long, pale yellow, densely imbricated with serrated margin, 4th segment base 0.08 mm long, 0.02 mm broad, primary rhinarium ciliated, round at tip of base; processus terminalis (p.t.) 0.30 mm long, pale yellow, faintly imbricated; p.t. longer than base, 3.75 x base of last antennal segment; antenna shorter than, 0.73 x body.

Flagellar Formula

1 L/W = 7.00, 2 L/W = 6.5, 4L/W= 19, A = 10.83.

Rostrum (Figure 65)

0.56 mm long, 0.05 mm broad, dark yellow, extend upto hindcoxa of metathoracic leg, 5 segmented; 4th and 5th segment fused to form ultimate rostral segment (u.r.s.), u.r.s. short, 0.11 mm long, 0.04 mm broad, slight blunt at apex, u.r.s. shorter than h.t.2, 0.91 x h.t.2.

THORAX

0.60 mm long, 0.72 mm broad, dark yellow, smooth, sparsely hairy with acute apices; distinct lateral tubercles, midthoracic furca sessile; prothorax 0.18 mm long, 0.44 mm broad; mesothorax 0.22 mm long, 0.72 mm broad; metathorax 0.16 mm long, 0. 74 mm broad.

Hind Leg (Figure 68)

1.20 mm long, pale yellow; coxa stout, 0.13 mm long, 0.12 mm broad, dark yellow, sparsely hairy with incrassate apices, broad at base, tapered at apex; trochanter short, triangular 0.08 mm long, 0.06 mm broad, dark yellow, sparsely hairy with incrassate apices; femur 0.25 mm long, 0.04 mm broad, pale yellow with densely hairy; tibia 0.65 mm long, 0.02 mm broad, pale yellow with densely hairy; tarsus 0.12 mm long, 0.02 mm broad; 1st tarsal segment very short, 0.02 mm long, pale yellow, conical, F.T.C. with 3, 3, 3; 2nd tarsal segment 0.12 mm long, 0.02 mm broad, slightly curved, imbricated with curved two claws, h.t. 2 with short hairs at dorsolateral side with incrassate apices; h.t.2 longer than u.r.s., 1.09 x u.r.s.

ABDOMEN

0.90 mm long, 0.85 mm broad, dark yellow, smooth, narrow anteriorly, broad posteriorly, sparsely hairy with incrassate apices, longest hair on abdomen shorter than 0.75 x basal diameter of 3rd segment; 8 segmented, 1st segment 0.12 mm long, 0.60 mm broad; 2nd segment 0.12 mm long, 0.65 mm broad; 3rd segment 0.13 mm long, 0.70 mm broad; 4th segment 0.11 mm long, 0.80 mm broad; 6th segment 0.11 mm long, 0.80 mm broad; 7th segment 0.09 mm long, 0.55 mm broad; 8th segment 0.09 mm long, 0.35 mm broad; subgenital plate very short with 4 short hairs on posterior margin with incrassate apices; subanal plate semioval, dark yellow with faintly spinulose.

Siphunculus (Figure 69)

0.29 mm long, 0.08 mm broad, dark yellow, faintly imbricated with distinct apical flange, without hairs, broad at base, tapered at apex; siphunculus longer than cauda, 1.93 x cauda and shorter than body, 0.17 x body.

Cauda (Figure 70)

0.15 mm long, 0.16 mm broad, dark yellow, elongated, broad at base, slight pointed at apex, cauda sparsely hairy with incrassate apices; cauda shorter than siphunculus, 0.51 x siphunculus.

HOST PLANTS

Rui: *Calatrophis gigantia.*

DISCUSSION

This species runs close to *Acyrthosiphon picum* Harris by

1. Ultimate rostral segment shorter than h.t. 2.
2. 3rd segment of antenna with 3 secondary rhinaria.
3. Cauda 0.51 x siphunculus.

However, it differ from above species by following characters.

1. Antenna shorter than the body.
2. Lateral frontal tubercles on head well developed, frontal sinus slightly raised, U shaped.
3. Dorsal cephalic hairs with incrassate apices.
4. Subgenital plate with 4 short hairs on posterior margin with incrassate apices.
5. Proportion of siphunculus and cauda.

6. Processus terminalis, 3.75 x base of last antennal segment.
7. Flagellar formula: 1 L/W = 7.00, 2 L/W = 6.5, 4L/W= 19, A = 10.83.

CONTROL MEASURES

Natural Control

Ladybird beetles: *Brumoides* sp., *Menochilus* sp. feed on aphids.

Chemical Control

Spray: 0.1 per cent carbaryl; 0.05 per cent malathion; 0.05 per cent Azadirachtin.

ACYRTHOSIPHON VATANI SP. NOV.

FEMALE: APTERA (Figures 71–75)

1.66 mm long, 0.88 mm broad, dark greenish yellow; head 0.20 mm long, 0.38 mm broad; antenna 6 segmented, 1.92 mm long, longer than body; thorax 0.62 mm long, 0.68 mm broad; abdomen 0.86 mm long, 0.88 mm broad, dark greenish yellow; siphunculus 0.45 mm long, 0.05 mm broad, pale green; cauda short, 0.18 mm long, 0.10 mm broad, pale green.

HEAD (Figure 71)

0.20 mm long, 0.38 mm broad, broader than long, smooth, dark greenish yellow; lateral frontal tubercles well developed, inner margin become shortly diverging type, frons with straight median prominence; dorsal cephalic hairs short, sparsely with incrassate apices; eyes lateral without triommatidia, ocular space 0.08 mm, interocular space 0.32 mm.

Antenna (Figure 73)

1.92 mm long, pale green, faintly imbricated, 6 segmented, longer than body; scape stout, 0.12 mm long, 0.10 mm board, dark green, sparsely hairy; pedicel short, straight, 0.08 mm long, 0.06 mm broad, dark green, sparsely hairy with incrassate apices; flagellum 4 segmented, distinctly imbricated, sparsely hairy with incrassate apices; 1st segment 0.32 mm long, 0.05 mm broad, with 5 secondary rhinaria at base; 2nd segment 0.29 mm long, 0.05 mm broad; 3rd segment 0.30 mm long, 0.03 mm broad; 4th segment 0.78 mm long, 0.02 mm broad, pale green, 4th segment base 0.15 mm long, 0.03 mm

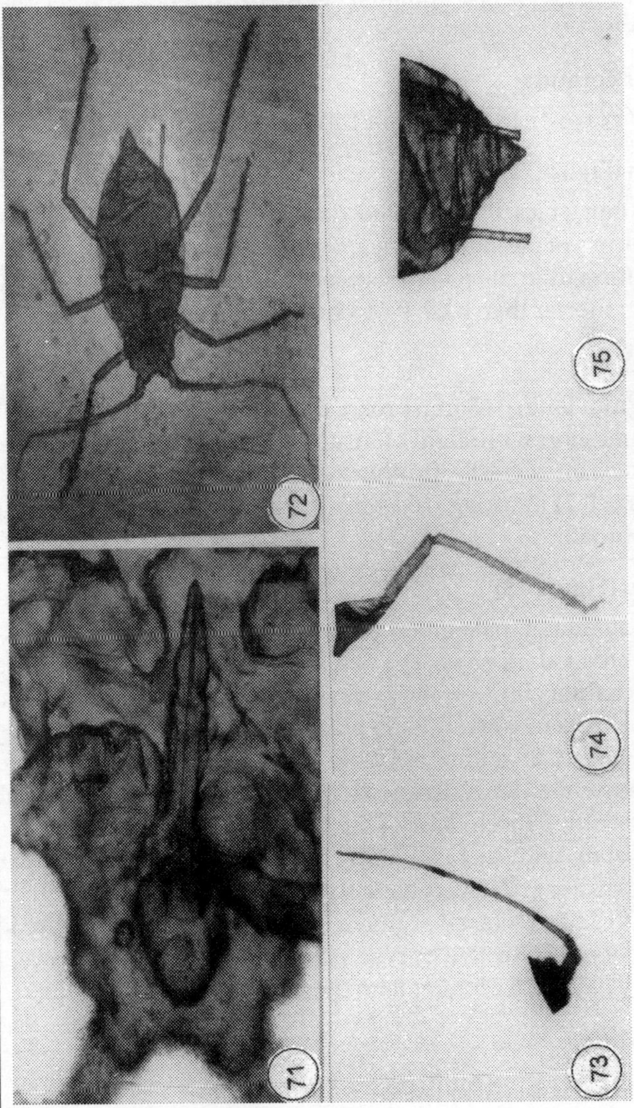

Figures 71-75: *Acyrthosiphon vatani* sp.nov. (Female Aptera)
Figure 71: Head and Rostrum; Figure 72: Whole Mount; Figure 73: Antenna; Figure 74: Hind Leg; Figure 75: Siphunculi and Cauda

broad, ciliated primary rhinarium at tip of base; processus terminalis (p.t.) 0.63 mm long, pale green, faintly imbricated, p.t. longer than base, 4.4 x base of last antennal segment; antenna shorter than, 1.56 x body.

Flagellar Formula

1 L/W = 6.4, 2 L/W = 5.8, 4 L/W = 26.0, A = 12.73.

Rostrum (Figure 71)

0.35 mm long, 0.05 mm broad, dark green, sparsely hairy, extend upto midcoxa of mesothoracic leg, 5 segmented; 4th and 5th segments fused to form ultimate rostral segment (u.r.s.), u.r.s. very short, 0.08 mm long, shorter than h.t.2, 0.44 x h.t. 2.

THORAX

0.62 mm long, 0.68 mm broad, dark greenish yellow, sparsely hairy, hairs short with slight blunt apices; distinct lateral tubercles, midthoracic furca sessile; prothorax 0.20 mm long, 0.42 mm broad; mesothorax 0.21 mm long, 0.60 mm broad; metathorax 0.21 mm long, 0.68 mm broad.

Hind Leg (Figure 73)

1.72 mm long, pale green, smooth; coxa stout, 0.12 mm long, 0.12 mm broad, dark green, sparsely hairy, broad at base, narrow at apex; trochanter short, straight, triangular, 0.05 mm long, 0.08 mm broad, dark green, sparsely hairy with incrassate apices; femur long, 0.50 mm long, 0.08 mm broad, pale green, smooth, slight curved, broad at apex; tibia long, narrow, 0.95 mm long, 0.05 mm broad, pale green, smooth, densely hairy with primary and secondary hairs dense at apex with incrassate apices; tarsus 0.18 mm long, pale green; 1st tarsal segment very short 0.02 mm long, pale green, conical, F.T.C. with 3,3,3; 2nd tarsal segment 0.18 mm long, 0.02 mm broad, slightly curved with two curved claws, h.t. 2 with 4 primary, 6 secondary hairs; h.t. 2 longer than u.r.s, 1.5 x u.r.s.

ABDOMEN

0.86 mm long, 0.88 mm broad, dark greenish yellow, smooth, broad anteriorly, tapered posteriorly, dorsal hairs very short, sparsely with incrassate apices, abdomen 8 segmented; 1st segment 0.10 mm long, 0.60 mm broad; 2nd segment 0.12 mm long, 0.68 mm broad; 3rd

segment 0.12 mm long, 0.88 mm broad; 4th segment 0.12 mm long, 0.70 mm broad, 5th segment 0.10 mm long, 0.60 mm broad; 6th segment 0.10 long, 0.41 mm broad; 7th segment 0.09 mm long, 0.28 mm broad; 8th segment short 0.06 mm long, 0.15 mm broad; subgenital plate with very short hair on hind margin with blunt apices, 0.50 x basal diameter of 3rd segment.

Siphunculus (Figure 75)

0.45 mm long, 0.05 mm broad, pale green, cylindrical faintly imbricated with distinct apical flange; siphunculus longer than cauda, 2.5 x cauda and shorter than body, 0.30 x body.

Cauda (Figure 75)

0.18 mm long, 0.10 mm broad, elongated, pale green, sparsely hairy with acute apices; cauda shorter than siphunculus, 0.35 x siphunculus.

HOST PLANTS

Pea; *Pisum sativum* Linn.

DISCUSSION

This species runs close to *Acrythosiphon pisum* Harris by

1. Ultimate rostral segment shorter than h.t. 2.
2. Antenna 6 segmented.

However, it differs from above species by having following characters.

1. Dorsal cephalic hairs with incrassate apices.
2. Antenna longer than body.
3. Cauda, 0.35 x siphunculi.
4. Lateral frontal tubercles on head well developed become shortly diverging, frons with straight median prominence.
5. Processus terminalis, 4.4 x base of last antennal segment.
6. Siphunculi, 0.30 x body.
7. Flagellar formula: 1 L/W = 6.4, 2 L/W = 5.8, 4 L/W = 26.0, A = 12.73.

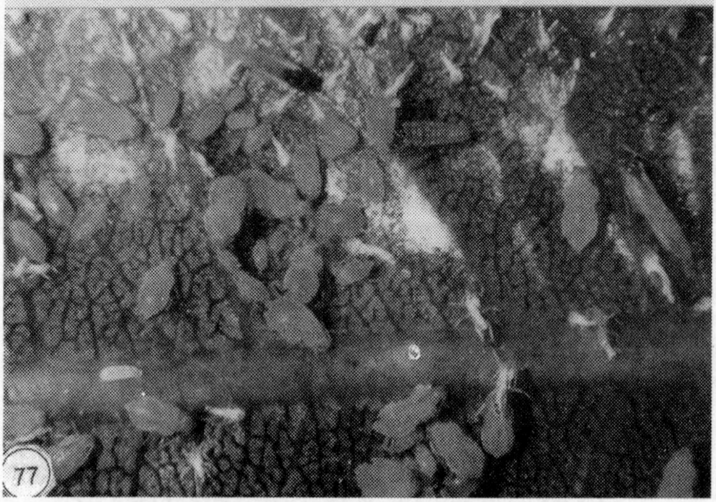

Figure 76: ***Acyrthosiphon calatropi*** **sp.nov.**
Figure 77: ***Acyrthosiphon vatani*** **sp.nov.**

CONTROL MEASURES

Natural Control

Ladybird beetles: *Coccinella* sp., and a lace wing, *Crysoperla carnea* cause mortalities in aphids.

Chemical Control

Spray: 0.05 per cent malathion; 0.1 per cent carbaryl; 0.03 per cent DDVP.

GENUS *BREVICORYNE* VAN DER GOOT

The genus *Brevicoryne* was erected by van der Goot in 1915 with the help of type species *Aphis brassicae* Linnaeus. The genus is cosmopolitan in distribution. It shows following characteristics.

1. Head smooth or slightly wrinkled and pigmented, weakly developed lateral frontal tubercles, median frontal tubercles higher than lateral frontal tubercles, dorsal cephalic hair long with acute to subacute apices.
2. Antenna 6 segmented, about 0.45–0.65 x body, 1st segment faintly imbricated, shorter than wide, 2nd segment as long as 1st segment, faintly imbricated; flagellum strongly imbricated, apterae without secondary rhinaria while alatae numerous rhinaria scattered on 3rd segment, primary rhinaria ciliated, flagellar hairs sparse, short with acute to acuminate apices; about 2.0–4.0 x base of last antennal segment.
3. Ultimate rostral segment reaches upto midcoxa of mesothoracic leg, normal, much shorter than h.t. 2 with 4 secondary hairs.
4. Prothorax in apterae with marginal brown patches and alatae with transverse pigmented bands.
5. Midthoracic furca with separate arms.
6. Legs smooth/with hairs having acute to subacute apices, faintly imbricated, tarsi with normal imbrications, F.T.C. 3,3,7.
7. Wing venation normal, all veins slightly to distinctly bordered brown, nymphs with hind tibiae smooth.
8. Abdominal dorsum in apterae with pleural pigmentation, irregular shaped patched on 1–7 segments, segment 8 with

spinopleural segmental band and scattered muscle plates on antisiphuncular segments, alatae with spinopleural segmental bar on each segment; dorsal hairs of moderate size, thick with blunt apices in apterae and acuminate apices in alatae placed on slightly raised bases.

9. Siphunculus about 0.05–0.08 x body, barrel shaped, in alatae basal portion much constricted, wrinkled, faintly imbricated, almost without any apical flange.
10. Cauda dark, triangular about 1.10–1.30 x siphunculus, bears 5-7 hairs become acute to acuminate apices.
11. Subgenital plate broadly oval with 10 hairs in 2 groups posteriorly and 2 hairs anteriorly.

Bodenheimer and Swirski (1957) considered *Brevicoryne* as a subgenus *Hyadaphis*. Eastop (1966) stated that it was closely related to *Lipaphis* and perhaps with this best regarded as a subgenus of *Hyadaphis* but *Lipaphis* and *Hyadaphis* are very closely related genera due to dorsal abdominal pigmentation is quite distinctive and brownish bordering of the wing veins of *Brevicoryne* is not found in *Hyadaphis*. *Brevicoryne* and *Hyadaphis* are, therefore, regarded here as distinct genera. From India, only one species has been reported under this genus.

BREVICORYNE CAPSICI SP. NOV.

FEMALE: APTERA (Figures 78–83)

1.89 mm long, 1.20 mm broad, pale green, globular; head 0.11 mm long 0.38 mm broad; antenna 6 segmented, 1.45 mm long, shorter than body; thorax 0.52 mm long, 0.99 mm broad; abdomen 1.30 mm long, 1.20 mm broad, pale green; siphunculus 0.17 mm long, 0.04 mm broad, pale green; cauda 0.18 mm long, 0.10 mm broad, dark green.

HEAD (Figure 78)

0.11 mm long, 0.38 mm broad, pale green, smooth, slight pigmented, lateral frontal tubercles weakly developed become broadly diverging type, well developed median frontal prominence; dorsal cephalic hairs longer than ventral hairs with acute apices; eyes lateral with distinct triommatidia, ocular space 0.10 mm, interocular space 0.32 mm.

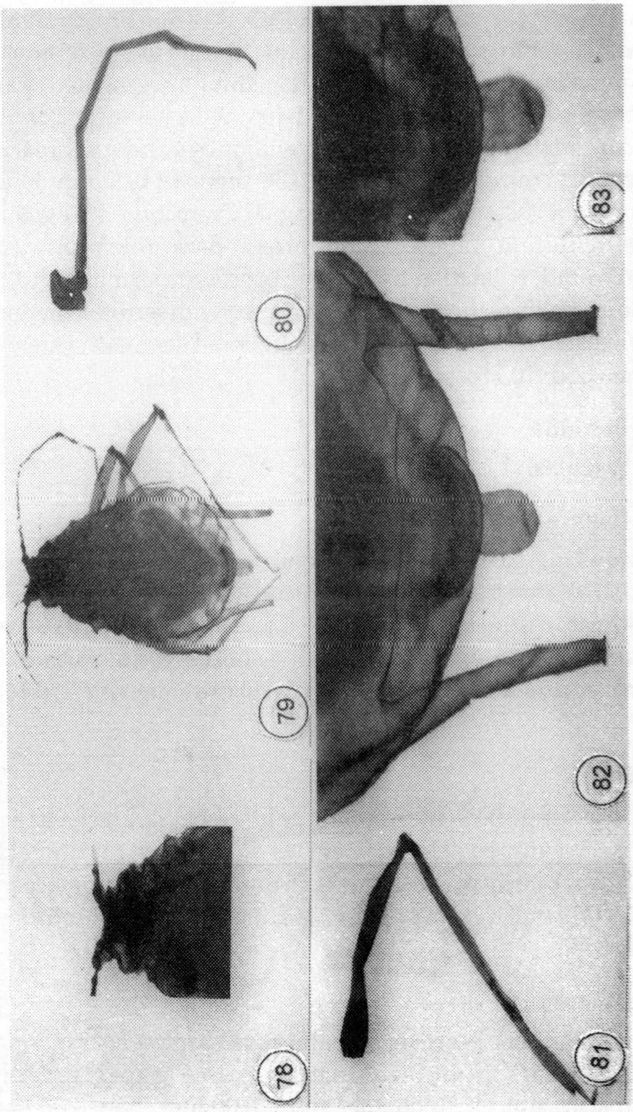

Figures 78-83: *Brevicoryne capsici* sp.nov. (Female Aptera)
Figure 78: Head and Rostrum; Figure 79: Whole Mount; Figure 80: Antenna; Figure 81: Hind Leg; Figure 82: Siphunculi; Figure 83: Cauda

Antenna (Figure 80)

1.45 mm long, pale green, distinctly imbricated, 6 segmented, shorter than body; scape stout, 0.08 mm long, 0.10 mm broad, faintly imbricated, shorter than wide; pedicel short 0.08 mm long, 0.08 broad, faintly imbricated, as long as scape; flagellum 4 segmented, pale green, densely imbricated, sparsely hairy with acute apices; 1st segment 0.32 mm long, 0.03 mm broad without secondary rhinaria; 2nd segment 0.30 mm long, 0.03 broad; 3rd segment 0.22 mm long, 0.02 mm broad; 4th segment 0.48 mm long, 0.02 mm long, 4th segment base short, 0.10 mm long, 0.02 mm broad, dark green, primary rhinarium round, ciliated at tip of base; processus terminalis (p.t.) 0.38 mm long, pale green, distinctly imbricated with serrated margin; p.t. longer than base, 3.8 x base of last antennal segment; antenna shorter than body, 0.76 x body.

Flagellar Formula

1 L/W =10.66, 2 L/W =10, 4 L/W = 24, A =14.88.

Rostrum (Figure 78)

0.35 mm long, 0.04 mm broad, smooth, pale green, extends upto hindcoxa of metathoracic leg, 5 segmented; 4th and 5th segment fused to form ultimate rostral segment (u.r.s.), u.r.s. much shorter, 0.08 mm long, 0.04 mm broad, acute at tip with 4 short secondary hairs, 2 long primary hairs with acute apices; u.r.s shorter than h.t. 2, 0.66 x h.t.2.

THORAX

0.52 mm long, 0.99 mm broad, pale green, sparsely hairy, marginal and scattered sclerotic patches on dorsum, midthoracic furca with separate arm; prothorax 0.15 mm long, 0.52 mm broad; mesothorax 0.19 mm long, 0.82 mm broad; metathorax 0.17mm long, 0.99 mm broad.

Hind Leg (Figure 81)

1.10 mm long, pale green, smooth; coxa stout, 0.12 mm long, 0.10 mm broad, dark green, broad at base, narrow at apex, sparsely hairy with acute apices; trochanter short, 0.06 mm long, 0.06 mm broad, dark green, sparsely hairy with acute apices; femur 0.18 mm long, 0.06 mm broad, pale green, slightly curved, broad at apex with short hairs with acute apices; tibia long, narrow, 0.65 mm long, 0.04 mm broad, pale green, smooth with normal imbrications, tibia sparsely hairy with acute apices; tarsus 0.12 mm long, pale green, 1st tarsal segment short, 0.02 mm long, pale green, F.T.C. with 3,3,3; 2nd tarsal segment 0.12 mm long, 0.02 mm broad, slightly curved, normally imbricated with curved two claws; h.t.2 longer than u.r.s., 1.50 x u.r.s.

ABDOMEN

1.30 mm long, 1.20 mm broad, pale green, smooth, sparsely hairy, short hairs with acute apices, 8 segmented; lateral abdominal tubercles indistinct on 1-7 segments; 1st segment 0.14 mm long, 0.95 mm broad; 2nd segment 0.16 mm long, 1.05 mm broad; 3rd segment 0.18 mm long, 1.10 mm broad; 4th segment 0.20 mm long, 1.20 mm broad; 5th segment 0.15 mm long, 1.15 mm broad; 6th segment 0.15 mm long, 0.85 mm broad; 7th segment 0.12 mm long, 0.65 mm broad; 8th segment 0.12 mm long 0.40 mm broad; subanal plate oval, dark brown, densely spinulose with 4 primary, 4 secondary hairs with acuminate apices.

Siphunculus (Figure 82)

0.17 mm long, 0.04 mm broad, pale green, barrel shaped, faintly imbricated without apical flange; siphunculus shorter than cauda, 0.94 x cauda and shorter than body, 0.08 x body.

Cauda (Figure 83)

0.18 mm long, 0.10 mm broad, dark green, dumbel shaped, broad at base, with slight middle constriction and blunt at apex; cauda with 7 short hairs with acute apices; cauda longer than siphunculus, 1.05 x siphunculus.

HOST PLANTS

Chili: *Capsicum annum* Linn.

DISCUSSION

This species runs close to *Brevicoryne brassicae* Linnaeus by

1. Ultimate rostral segment shorter than h.t. 2.
2. Siphunculi shorter than cauda.
3. Processus terminalis, 3.8 x base of last antennal segment.

However, it differs from above species by having following characters:

Figure 84: ***Brevicoryne capsici*** **sp.nov.**
Figure 85: Colony (Enlarged) of ***Brevicoryne capsici*** **sp.nov.**

1. Lateral frontal tubercles on head weakly developed become broadly diverging type, frons with well developed median prominence.
2. Cauda dumbel shaped, middle constriction with blunt apex.
3. Subanal plate oval with 4 primary, 4 secondary hairs with acuminate apices.
4. Dorsal abdominal hairs short with acute apices.
5. Proportion of siphuncules to cauda.
6. Flagellar formula: 1 L/W =10.66, 2 L/W =10, 4 L/W = 24, A =14.88.

CONTROL MEASURES

Natural Control

Ladybird beetles: *Menochilus* sp. and hover fly feed on aphid populations.

Chemical Control

Spray: 0.03 per cent malathion; 0.03 per cent Azadiractin.

GENUS *LIPAPHIS* MORDVILKO

The genus *Lipaphis* was erected by Mordvilko in 1928. It is cosmopolitan genus and widely distributed in India. It has been reported from Arunachal Pradesh, Assam, Bihar, Himachal Pradesh, Manipur, Megalaya, Nagaland, Rajasthan, Sikkim, South India, Tripura, Uttar Pradesh, West Bengal, Bhutan and Nepal. The type species used for erecting the new genus was *Aphis erysimi* Kaltenbach. The genus shows following characteristics.

1. Head in apterae shows spinulose both surfaces, in alatae only ventral spinulose, distinct lateral frontal tubercles, frons straight, scabrous.
2. Antenna 6 segmented, slightly longer than body, 1st segment imbricated, 2nd segment only ventrally scabrous and flagellum imbricated except 3rd segment, secondary rhinaria absent apterae while in alatae numerous, circular, protuberant on 3rd and 4th segment, primary rhinarium ciliated, p.t. much longer than base of last antennal segment.

3. Rostrum long, extend upto hind coxa of metathoracic segment, u.r.s. longer than h.t.2, with secondary hairs.
4. Midthoracic furca with short stalk.
5. In apterae abdomen dorsum pale, usually with distinct reticulated pattern, post siphuncular area become scabrous; in alatae with scabrous area on segment 2-5 with median pigmented patches on segment 3-5, broad postsiphuncular patch on segment with pigmented transverse band; dorsal abdominal hairs short and shorter on anterior tergite and on 8th tergite long with subacute apices.
6. Siphunculus cylindrical, reticulated, 0.21 x 0.27 x body, imbricated, dark, without district flange.
7. Cauda dark, short, conical with bulbous base, bearing 4-5 hairs and about 0.24–0.26 x siphunculi.
8. Subgenital plate broad, rounded posterior margin with 5 blunt hairs in 2 groups and 4 longer hairs with acuminate apices.
9. Spiracles on segments 6–7 placed more apart than those on segment 5–6 abdomen.
10. Femur possess spinulose ventrally, tibia smooth, tarsi with normal imbrications, F.T.C. is 3, 3, 3.
11. Wing venation of forewing and hind wings normal, cubitus and anal vein of forewing thicker, faintly bordered brown, costal area with brown spots.

According to Eastop (1966) there is confusion with *Lipaphis* and *Hyadophis*, but can be cleared by strongly elevate siphunculi and somewhat weakly developed flange in *Hyadophis*, in contrast to slender siphunculi and well developed flange found in *Lipaphis*. From India, only one species of this genus has been reported.

LIPAPHIS ERYSIMI KALTENBACH

FEMALE: APTERA (Figures 86–91)

1.78 mm long, 0.84 mm broad, dark brown, head 0.15 mm long, 0.35 mm broad; antenna 6 segmented, 1.21 mm long, shorter than body; thorax 0.62 mm long, 0.48 mm broad; abdomen 1.10 mm long, 0.84 mm broad, slight pale brown; siphunculus 0.24 mm long, 0.04

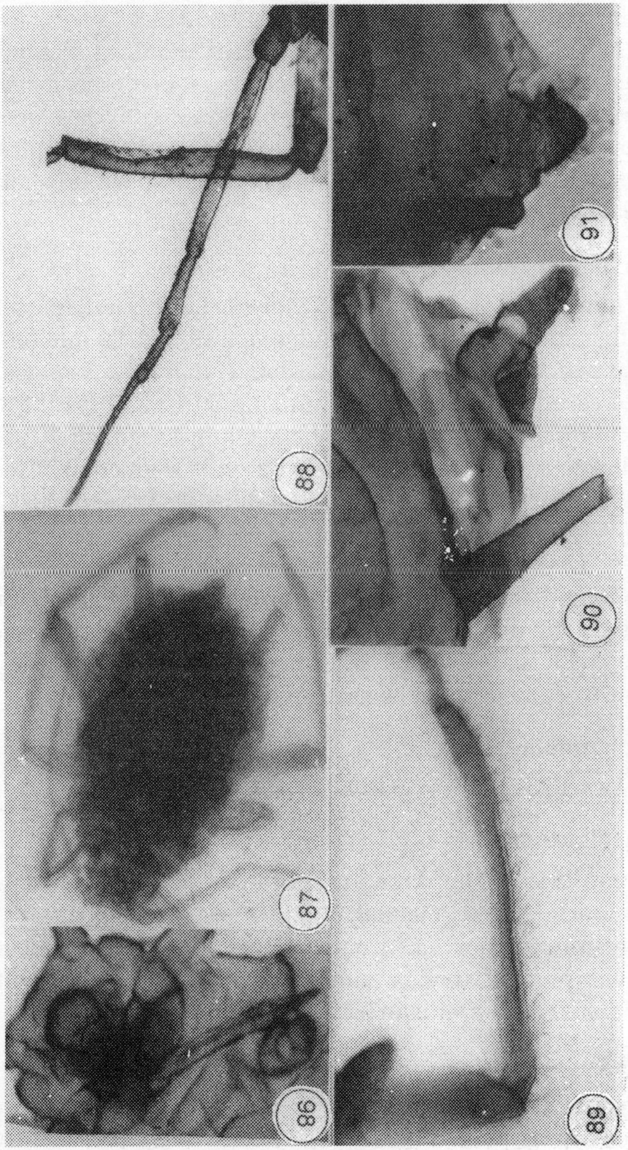

Figures 86-91: *Lipaphis erysimi* (Female Aptera)
Figure 86: Head and Rostrum; Figure 87: Whole Mount; Figure 88: Antenna; Figure 89: Hind Leg; Figure 90: Siphunculi; Figure 91: Cauda

mm broad, dark brown, subcylindrical; cauda 0.08 mm long, 0.04mm broad, dark brown.

HEAD (Figure 86)

0.15 mm long, 0.35 mm broad, dark brown, wrinkled, scabrous, pigmented; lateral frontal tubercles weakly developed, inner margin diverging, frons with median frontal prominence slightly developed; dorsal cephalic hairs very short, sparsely with incrassate apices; eyes lateral with distinct triommatidia, ocular space 0.10 mm, interocular space 0.30 mm.

Antenna (Figure 88)

1.21 mm long, pale brown with distinctly imbricated, sparsely hairy with acute apices, 6 segmented; scape stout, 0.06 mm long, 0.08 mm broad, dark brown, scabrous, longer than broad, sparsely hairy; pedicel short, straight, 0.06 mm long, 0.04 mm broad, dark brown, scabrous with sparsely hairy; flagellum 4 segmented, pale brown, distinctly imbricated, sparsely hairy; 1st segment 0.27 mm long, 0.02 mm broad, without secondary rhinaria; 2nd segment 0.26 mm long, 0.02 mm broad; 3rd segment 0.21 mm long, 0.02 mm broad; 4th segment 0.33 mm long, pale brown, imbricated, 4th segment base 0.09 mm long, 0.02 mm broad, dark brown, primary rhinarium ciliated, flat at tip of base; processus terminalis (p.t.) 0.24 mm long, imbricated with sparsely hairy, p.t. longer than base, 2.65 x base of last antennal segment; antenna shorter than body, 0.69 X body.

Flagellar Formula

1 L/W = 13.5, 2 L/W= 13.0, 4 L/W= 16.5, A = 14.33.

Rostrum (Figure 86)

0.38 mm long, 0.04 mm broad, dark brown, extend up to midcoxa of mesothoracic leg, 5 segmented; 4th and 5th segment fused to form ultimate rostral segment (u.r.s.), u.r.s. 0.09 mm long, 0.03 mm broad, normal with pointed at apex with 4 short, secondary hairs with acute apices; u.r.s. shorter than h.t.2, 0.75 x h.t.2.

THORAX

0.62 mm long, 0.48 mm broad, dark brown, rugose, scabrous, diffuse brown segmented bars sparsely hairy, midthoracic furca sessile; prothorax 0.16 mm long, 0.36 mm broad, separate from head and mesothorax; mesothorax 0.30 mm long, 0.48 mm broad, fused with metathorax; metathorax 0.14 mm long, 0.44 mm broad.

Hind Leg (Figure 89)

1.08 mm long, slight pale brown, imbricated; coxa stout, 0.10 mm long, 0.12 mm broad, imbricated, slightly dark brown, broad at base, tapered at apex, sparsely hairy with blunt apices; trochanter short, straight, 0.06 mm long, 0.07 mm broad, slight dark brown, smooth, sparsely hairy with blunt apices; femora 0.22 mm long, 0.06 mm broad, ventrally scabrous, slight curved at apex, remaining femora smooth with sparsely hairy; tibia long, cylindrical, 0.60 mm long, 0.05 mm broad, faintly imbricated at base, rest smooth with dense primary and secondary hairs with blunt apices; tarsus 0.12 mm long, 0.02 mm broad; 1st tarsal segment short, 0.02 mm long, conical, pale brown, F.T.C. with 3, 3, 3 fine hairs; 2nd tarsal segment (h.t.2) 0.12 mm long, slightly curved with curved two claws, sparsely hairy with blunt apices; h.t. 2 longer than u.r.s., 1.3 x u.r.s.

ABDOMEN

1.10 mm long, 0.84 mm broad; dark brown, oval, dorsum weakly wrinkled with faint reticulations on anterior segments, sparsely hairy with slight blunt apices, 8 segmented; lateral abdominal tubercles indistinct on 3–5 segments; 1st segment 0.10 mm long, 0.38 mm broad; 2nd segment 0.10 mm long, 0.44 mm broad; 3rd segment 0.12 mm long, 0.59 mm broad; 4th segment 0.13 mm long, 0.70 mm broad; 5th segment 0.14 mm long, 0.84 mm broad; 6th segment 0.12 mm long, 0.72 mm broad; 7th segment 0.10 mm long, 0.55 mm broad; 8th segment 0.20 mm long, 0.26 mm broad; subanal plate normal, oval; subgenital plate normal, dark brown, 8 short, hairs with blunt apices.

Siphunculus (Figure 90)

0.24 mm long, 0.04 mm broad, subcylindrical, dark brown, imbricated, slight constricted at middle with indistinct apical flange at apex; siphunculus longer than cauda, 3.0 x cauda and shorter than body, 0.13 x body.

Cauda (Figure 91)

0.08 mm long, 0.04 broad, elongated, tapered at apex with 4 long basal hairs, 2 short apical hairs with blunt apices; cauda shorter than siphunculus, 0.33 x siphunculus.

HOST PLANTS

Pigeon pea: *Cajanus cajan* Mill; Subabul–*Leucaena;* Khair–*Acacia catechu;*Khinja–*Terminalia paniculata;* Mustard–*Brassica compestris;* Cabbage–*Brassica capitata;* Cauliflower–*Brassica botrytis.*

CONTROL MEASURES

Natural Control

Ladybird beetles: *Brumoides* sp., *Crysopa* sp. suppress the aphid population.

Chemical Control

Spray: 0.02 per cent phosphamidon; 0.05 per cent Rogor

Dusting: 5 per cent malathion.

GENUS *PENTALONIA* COQUEREL

The genus *Pentalonia* was erected by Coquerel in 1959. The genus is cosmopolitan and distributed throughout India.

Type species: *Pentalonia nigronervosa* Coquerel.

The genus shows following characteristics.

1. Head pigmented, densely spinulose on both surfaces, lateral frontal tubercles low but distinct, with diverging type, median frontal prominence is lower in height than lateral frontal tubercles.
2. Antenna 6 segmented, shorter to longer than body, 1st segment almost as long as wide with spinulose imbrications, 2nd segment heavily scabrous, slightly shorter than 1st segment, flagellum densely imbricated, apterae without secondary rhinaria while alatae with circular non protuberant secondary rhinaria on 3rd, 4th and 5th segment, primary rhinaria ciliated, flagellar hairs short with blunt apices, p.t. much longer than base of 6th segment.
3. Rostrum extends little beyond hindcoxa, u.r.s. normal, longer than h.t.2 with 2-6 secondary hairs.
4. Midthoracic furca with separate arms.
5. Abdominal dorsum with polygonal reticulations particularly on antisiphuncular segments and with

transverse spinular striae on postsiphuncular segments, in alatae this pattern faintly indicated spinular patches on segments 2-7, dorsal hairs short with expanded apices, placed on raised sockets.

6. Siphunculus compressed at middle and distally slightly attenuated, without distinct apical flange, about 0.13–0.23 x body with dense spinulose imbrications.
7. Cauda short, dusky with bulbous base and blunt apices, about 0.35–0.42 x siphunculus with 4 -6 hairs.
8. Subgenital plate normal with very short, blunt hairs on hind margin and 4 slightly longer hairs on anterior margin.
9. Coxa, trochanter and femora densely spinulose but tibia smooth or sometimes faintly imbricated, hairs mostly short and blunt apices, tarsal imbrications normal, F.T.C. 3, 3, 2.
10. Wing venation typical, radial sector fusing with anterior branch of media in forewing, hindwing with one obligue vein, all veins strongly bordered brown.

From India, only one species has been reported under this genus.

PENTALONIA KELI SP. NOV.

FEMALE: APTERA (Figures 92–96)

1.60 mm long, 0.90 mm broad, dark brown; head 0.20 mm long, 0.39 mm broad; antenna 6 segmented, 1.45 mm long, shorter than body; thorax 0.46 mm long, 0.82 mm broad; abdomen 0.90 mm long, 0.90 mm broad; siphunculus 0.28 mm long, 0.06 mm broad, dark brown; cauda short, 0.08 mm long, 0.06 mm broad.

HEAD (Figure 92)

0.20 mm long, 0.39 mm broad, dark brown, pigmented, spinulose; lateral frontal tubercles weakly developed, diverging, frons with weakly developed median frontal prominence forming zigzag shape; dorsal cephalic hairs short with fan shaped apices, ventral hairs slightly long with blunt apices; eyes lateral with distinct triommatidia, ocular space 0.10 mm, interocular space 0.33mm.

Antenna (Figure 94)

1.45 mm long, pale brown, densely imbricated, scabrous, 6 segmented, shorter than body; scape 0.08 mm long, 0.08 mm broad,

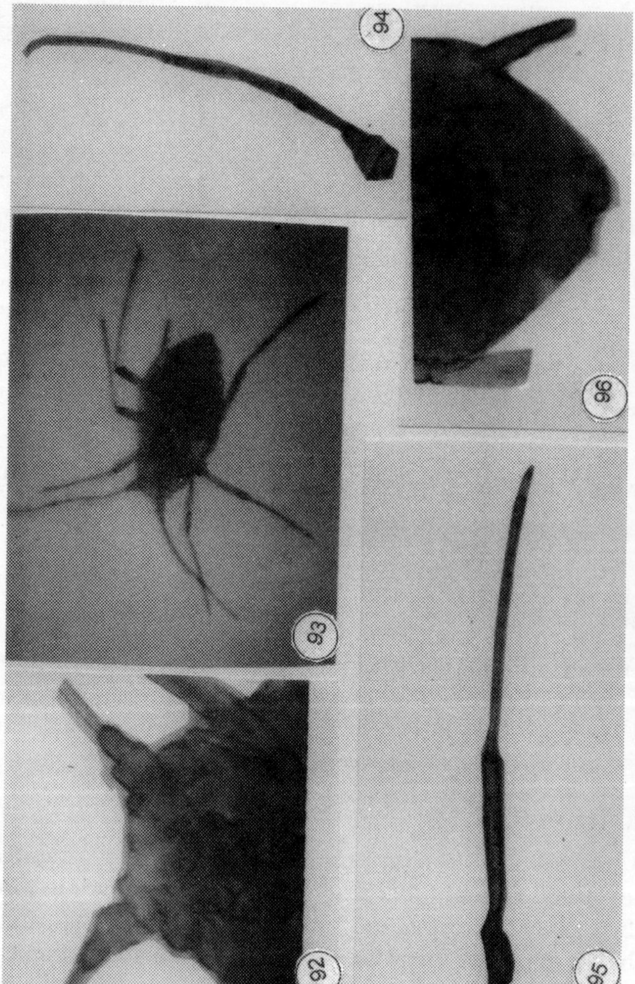

Figures 92-96: *Pentalonia keli* sp.nov. (Female Aptera)
Figure 92: Head and Rostrum; Figure 93: Whole Mount; Figure 94: Antenna; Figure 95: Hind Leg; Figure 96: Siphunculi and Cauda

dark brown, as long as broad, scabrous, imbricated; pedicel 0.07 mm long, 0.06 mm broad, dark brown, heavily scabrous, imbricated; flagellum 4 segmented, pale brown, densely imbricated; 1st segment 0.29 mm long, 0.03 mm broad, without secondary rhinaria; 2nd segment 0.19 mm long, 0.02 mm broad; 3rd segment 0.18 mm long; 0.02 mm broad; 4th segment 0.64 mm long, 0.02 mm broad, pale brown, imbricated, 4th segment base short, 0.08 mm long, 0.02 mm broad, primary rhinarium ciliated, round at tip of base; processus terminalis (p.t.) 0.56 mm long, pale brown, densely imbricated with serrated margin; p.t. longer than base, 7.00 x base of last antennal segment; antenna shorter than body, 0.90 x body.

Flagellar Formula

1 L/W = 9.6, 2 L/W = 9.5, 4 L/W = 32.00, A= 17.05.

Rostrum (Figure 92)

0.52 mm long, 0.06 mm broad, pale brown, extend slight behind hindcoxa of metathoracic leg upto anterior part of abdomen, 5 segmented; 4th and 5th segment fused to form ultimate rostral segment (u.r.s.), u.r.s. 0.14 mm long, 0.05 mm broad, pale brown, broad at base, pointed apex, 8 secondary hairs with acute apices; u.r.s. longer than h.t.2, 1.75 x h.t.2.

THORAX

0.46 mm long, 0.82 mm broad, dark brown patches laterally with irregular transverse striations, broader than long, midthoracic furca with separate arms; prothorax 0.14 mm long, 0.46 mm broad with dark patches on lateral side; mesothorax 0.16 mm long, 0.66 mm broad; metathorax 0.14 mm long, 0.82 mm broad, with dark patches on lateral side.

Hind Leg (Figure 95)

1.25 mm long, dark brown, densely spinulose, sparsely hairy; coxa stout, 0.08 mm long, 0.10 mm broad, dark brown, densely spinulose, sparsely hairy, broad at base, tapered at apex; trochanter short, straight, 0.06 mm long, 0.08 mm broad, dark brown, densely spinulose, sparsely hairy; femur 0.38 mm long, 0.04 mm broad, densely spinulose, pale brown at base, dark green at apex, hairs short, thick with blunt apices; tibia long, narrow, 0.69 mm long, 0.03 mm broad, smooth, faintly imbricated, hairs short, thick with blunt apices; tarsus short, 0.08 mm long, 0.02 mm broad, pale brown, 1st tarsal segment very short, 0.02 mm long, pale brown, conical, F. T. C. with 3, 3, 2; 2nd tarsal segment (h.t.2) 0.08 mm long, slightly curved, normal imbrications with curved two claws, h.t.2 shorter than u.r.s., 0.57 x u.r.s.

ABDOMEN

0.90 mm long, 0.90 mm broad, dark brown, polygonal reticulations, transverse striae on post siphuncular segments, dorsal hairs short with expanded apices; 8 segmented; 1st segment 0.09 mm long, 0.86 mm broad; 2nd segment 0.12 mm long, 0.88 mm broad; 3rd segment 0.12 mm long, 0.90 mm broad; 4th segment 0.14 mm long, 0.90 mm broad; 5th segment 0.14 mm long, 0.80 mm broad; 6th segment 0.12 mm long, 0.55 mm broad; 7th segment 0.12 mm long, 0.40 mm broad; 8th segment 0.14 mm long, 0.30 mm broad; subgenital plate normal with very stout, blunt hairs on hind margin, 4 longer hairs on anterior margin; subanal plate oval, dark brown, very short hairs with blunt apices.

Siphunculus (Figure 96)

0.28 mm long, 0.06 mm broad, dark brown, densely spinulose and imbricated, without distinct apical flange at tip; siphunculus longer than cauda, 3.3x cauda and shorter than body, 0.17 x body.

Cauda (Figure 96)

0.08 mm long, 0.06 mm broad, short, dark brown, blunt at apex, 7 secondary hairs with blunt apices; cauda shorter than siphunculus, 0.28 x siphunculus.

HOST PLANTS

Banana: *Musa paradisiaca* Linn.

DISCUSSION

This species runs close to *Pentalonia nigronervosa* Coquerel by having

1. Processus terminalis much longer than base, 7.0 x base of last antennal segment.
2. Dorsal cephalic hairs short with fan shaped apices.
3. Subgenital plate normal, 4 longer hairs at anterior margin.

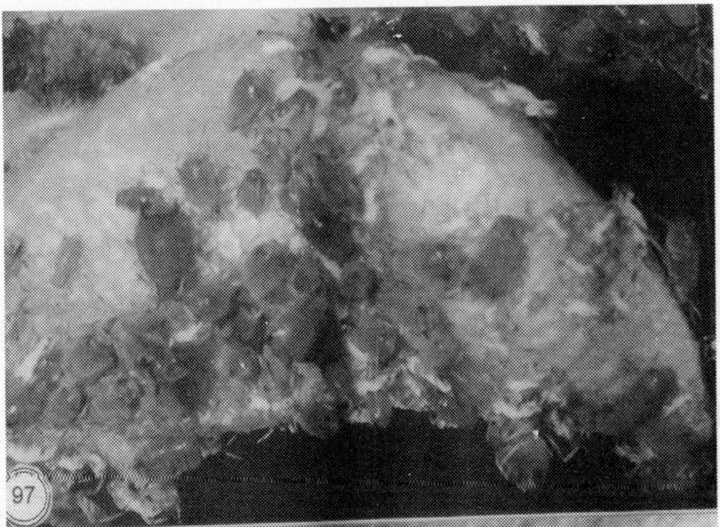

Figure 97: Colony of *Lipaphis erysimi*
Figure 98: Colony of *Pentalonia keli* sp.nov.

However, it differs from above species by having following characters.

1. Antenna shorter than body.
2. Ultimate rostral segment longer than h.t. 2.
3. Lateral frontal tubercles on head weakly developed, frons with weakly developed median frontal prominence forming zigzag shape.
4. Proportion of siphunculus to cauda.
5. Caudal 0.08 mm, shorter *i.e.* shorter than *P. nigronervosa.*
6. Caudal 7 secondary hairs with blunt apices.
7. Flagellar formula: 1 L/W = 9.6, 2 L/W = 9.5, 4 L/W = 32.00, A= 17.05.

CONTROL MEASURES

Natural Control

Ladybird beetles and lace wings feed on this aphid.

Chemical Control

Spray: 0.03 per cent DDVP/0.03 per cent phosphamidon/0.03 per cent malathion/0.03 per cent Azadirachtin; Apply carbofuran at 4.5 kg/ha.

GENUS *INDOMYZUS* GHOSH, GHOSH, RAYCHAUDHURI

The genus *Indomyzus* was erected by Ghosh, Ghosh and Raychaudhuri in 1971 with the help of type species *Indomyzus sensoriatus* Ghosh, Ghosh and Raychaudhuri. The genus is distributed throughout India. The genus shows following characters.

1. Head dark with small rounded warts, with prominent diverging lateral frontal tubercles, dorsal cephalic hairs long on strong sockets, long with blunt to incrassate apices.
2. Antenna 6 segmented, shorter than body, basal segments scabrous, flagellum imbricated, 3rd segment with round, protuberant secondary rhinaria nearly in row over entire length, sometime also on 4th segment, primary rhinaria protuberant without cilia, p.t. about 3.0–3.40 x base of last

segment, flagellar hairs long with blunt to acuminate apices.

3. Ultimate rostral segment normal, longer than h.t. 2. with 8-10 secondary hairs.
4. Head, prothorax and mesothorax distinct, metathorax fused with first 7 abdominal segments, prothorasic and mesothorasic tergites brown; tergites of fused thorasic and abdominal region also dark, sclerotic.
5. Abdominal dorsum smooth except area cauded to the siphunculi, spinulose, dorsal hairs long, thick, stiff with somewhat acuminate to bluntish apices, longest hair on anterior tergites about 1.60 x b. d. III, venter with distinct spinulose striae with laterally scabrous; 8th abdominal tergite also dark and sclerotic
6. Siphunculus dark, short, subcylindrical with apical flange, imbricated, 1.50–1.80 x cauda.
7. Dark bluntly conical cauda bearing 5 hairs.
8. Subanal plate with semicircular posterior margin.
9. Subgenital plate in usual shape with 16 short hairs, in 2 clusters on posterior margin and 2 long fine hairs anteriorly.
10. Femur smooth on margin, few spinules on surface near apex, femoral hairs with acuminate spices, tibiae smooth, 2nd tarsal segment with normal imbrications, F.T.C. 3,3,2.

The genus *Indomyzus* resembles with *Micromyzus* van der Goot by having secondary rhinaria on 3rd segment in apterae but differs by the nature of siphunculi, it can also be differentiated from *Paramyzus* Takahashi by presence of dark skelerotized body, siphunculus and cauda. From India, only one species has been reported under this genus.

INDOMYZUS GLARICAE SP. NOV.

FEMALE: APTERA (Figures 97–102)

1.45 mm long, 0.65 mm broad, dark brown; head 0.20 mm long, 0.28 mm broad; antenna 6 segmented, 0.95 mm long, shorter than body; thorax 0.50 mm long, 0.48 mm broad; abdomen 0.75 mm long,

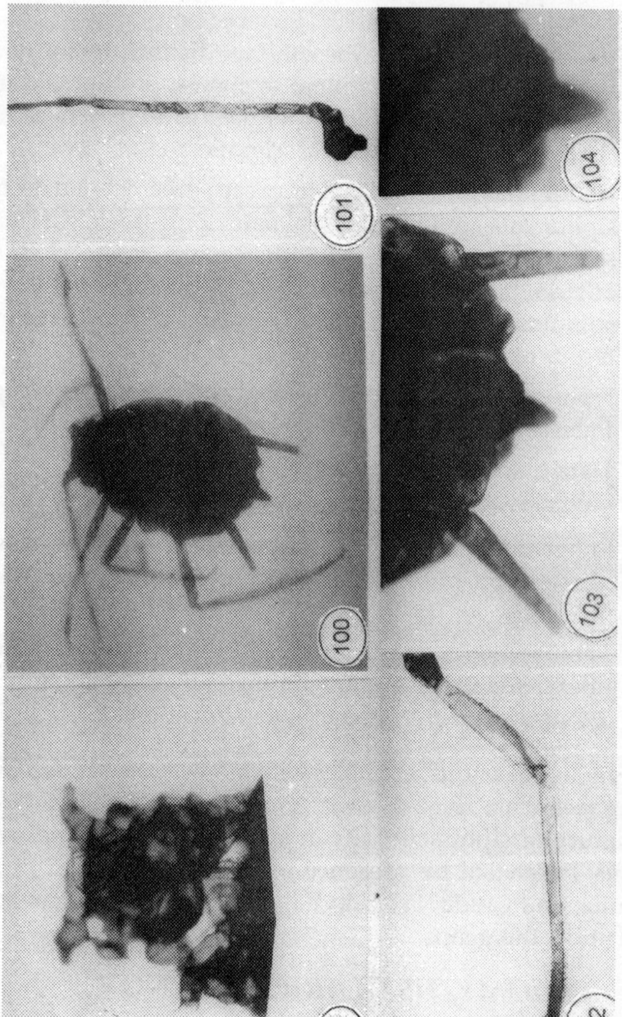

Figures 99-104: *Indomyzus glaricae* sp.nov. (Female Aptera)
Figure 99: Head and Rostrum; Figure 100: Whole Mount; Figure 101: Antenna;
Figure 102: Hind Leg; Figure 103: Siphunculi; Figure 104: Cauda

0.70 mm broad, dark brown; siphunculus short, 0.20 mm long, 0.04 mm broad, dark brown; cauda 0.08 mm long, 0.04 mm broad, dark brown.

HEAD (Figure 97)

0.20 mm long, 0.28 mm broad, dark brown, scabrous; lateral frontal tubercles well developed straightly diverging, frons with weakly developed median frontal prominence; dorsal, ventral cephalic hairs long with incrassate apices; eyes lateral with distinct triommatidia, ocular space 0.08 mm, interocular space 0.22 mm.

Antenna (Figure 99)

0.95 mm long, pale brown, imbricated, sparsely hairy, antenna 6 segmented, shorter than body; scape stout, 0.06 mm long, 0.08 mm broad, dark brown, scabrous, sparsely hairy with incrassate apices; pedicel short, 0.04 mm long, 0.04 mm broad, dark brown, scabrous, sparsely hairy with incrassate apices; flagellum 4 segmented, pale brown, faintly imbricated, hairs long with blunt apices; 1st segment 0.22 mm long, 0.03 mm broad with 4 secondary rhinaria, sparsely hairy, longest hair on 3rd segment longer than basal diameter; 2nd segment 0.20 mm long, 0.03 mm broad, secondary rhinaria absent; 3rd segment 0.16 mm long, 0.02 mm broad; 4th segment 0.34 mm long, 0.02 mm broad, pale brown, faintly imbricated, 4th segment base 0.08 mm long, 0.02 mm broad, sparsely hairy, imbricated, primary rhinarium protuberant without cilia at tip of base; processus terminalis (p.t.) 0.26 mm long, imbricated, dark brown, p.t. longer than base, 3.2 x base of last antennal segment; antenna shorter than body, 0.65 x body.

Flagellar Formula

1 L/W = 6.6, 2 L/W = 6.6, 4 L/W = 17, A=10.88.

Rostrum (Figure 97)

0.35 mm long, 0.04 mm broad, long, pale brown, extend upto hindcoxa of metathoracic leg, 5 segmented; 4th and 5th segment fused to form ultimate rostral segment (u.r.s.), u.r.s. 0.14 mm long, 0.04 mm broad, pale brown, u.r.s. normal with 12 secondary hairs with blunt apices; u.r.s. longer than h.t.2, 1.16 x h.t.2.

THORAX

0.50 mm long, 0.48 mm broad, dark brown with sclerotic plate, prothorax, mesothorax distinctly separate, metathorax fused with abdominal segments; prothorax 0.12 mm long, 0.34 mm broad with lateral tubercle; mesothorax 0.25 mm long, 0.45 mm broad; metathorax 0.22 mm long, 0.48 mm broad.

Hind Leg (Figure 100)

1.16 mm long, pale brown; coxa stout, 0.10 mm long, 0.08 mm broad, dark brown, smooth, without hairs, broad at base, narrow at apex; trochanter short, 0.04 mm long, 0.04 mm broad, dark brown, smooth, without hairs; femur straight, 0.36 mm long, 0.04 mm broad, smooth, few hairs near apex with acuminate apices; tibia narrow, 0.52 mm long, 0.03 mm broad, pale brown, densely hairy with acuminate apices; tarsus 0.12 mm long, 0.02 mm broad, pale brown, 1st tarsal segment very short, 0.02 mm long, pale brown, F.T. C. with 3, 3, 2; 2nd tarsal segment (h. t. 2) 0.12 mm long, pale brown with normal imbrications, slightly curved with curved two claws, h.t.2 shorter than u.r.s., 0.85 x u.r.s.

ABDOMEN

0.75 mm long, 0.70 mm broad, dark brown, smooth, spinulose at posteriorly with quadrangular patches at dorsolateral side, dorsal hairs long, thick, stiff with acuminate apices, longest hairs on anterior abdomen, about 2.00 x basal diameter of 3rd segment, 8 segmented; 1st segment 0.10 mm long, 0.46 mm broad; 2nd segment 0.11 mm long, 0.60 mm broad; 3rd segment 0.11 mm long, 0.64 mm broad; 4th segment 0.12 mm long, 0.70 mm broad; 5th segment 0.85 mm long, 0.65 mm broad; 6th segment 0.07 mm long, 0.55 mm broad; 7th segment 0.06 mm long, 0.40 mm broad; 8th segment 0.09 mm long, 0.20 mm broad with 6 long fine hairs with acute apices; subanal plate with semicircular posterior margin; subgenital plate usual shape with 16 hairs in 2 clusters on posterior margin and 4 long, fine hairs anteriorly.

Siphunculus (Figure 101)

0.20 mm long, 0.04 mm broad, dark brown, subcylindrical, slightly imbricated, smooth, without hairs with indistinct apical flange; siphunculi longer than cauda, 2.5 x cauda.

Cauda (Figure 102)

0.08 mm long, 0.04 mm broad, dark brown, elongated, knob like with 2 long stiff hairs and 8 short hairs; cauda shorter than siphunculus, 0.4 x siphunculus.

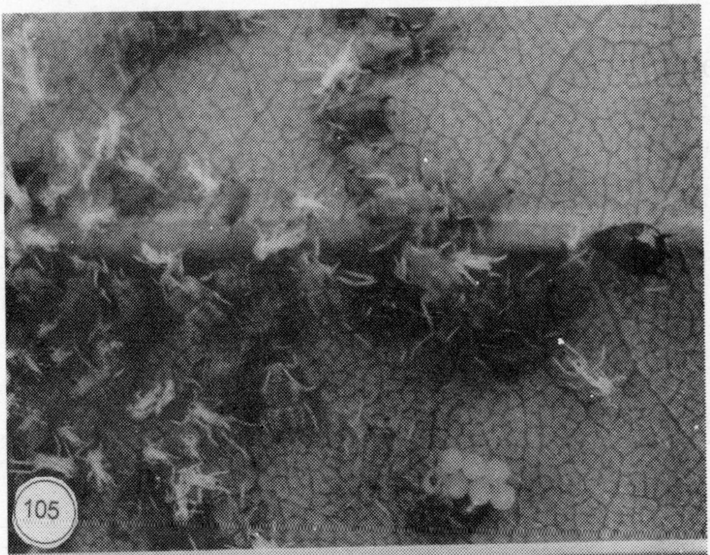

Figure 105: Colony of *Indomyzus glaricae* sp.nov. with Eggs of *Coccinellid beetle*
Figure 106: Colony of *Indomyzus glaricae*

HOST PLANTS

Giripushpa: *Gliricidia maculata* Syn.

DISCUSSION

This species runs close to *Indomyzus sensoriatus* Ghosh, Ghosh and Raychaudhuri by having

1. Ultimate rostral segment longer than h.t. 2.
2. Dorsal abdominal hairs long, thick, stiff with acuminate apices.
3. Longest hair on anterior tergite, 2.00 x basal diameter of 3rd segment.

However, differs from above species by following characters:

1. 8th tergite with 6 long hairs with acute apices.
2. Siphunculi with indistinct apical flange.
3. Subgenital plate with 4 long hairs with fine apices.
4. Ultimate rostral segment with 12 secondary hairs.
5. Lateral frontal tubercles on head well developed, straightly diverging, frons with weakly developed median frontal prominence.
6. Proportion of siphunculus to cauda.
7. Flagellar formula: 1L/W = 6.6, 2 L/W = 6.6, 4 L/W = 17, A=10.88.

CONTROL MEASURES

Natural Control

Ladybird beetles, lacewing, Syriphid feed on this pest.

Chemical Control

Spray: 0.02 per cent phosphamidon or 0.03% Azadirachtin.

GENUS *MYZUS* PASSERINI

Passerini (1860) erected the genus *Myzus* with the help of type species *Aphis cerasi* Fabricius. It is cosmopolitan genus. The genus shows following characters:

1. Head dorsally spinulose or warty, ventrally also spinulose, median prominence hardly or weekly developed, lateral

frontal tubercles moderately to well developed with inner margin converging to diverging.

2. Antenna 6 segmented, shorter to longer than body, 1st and 2nd segment scabrous, flagellum gradually more distinctly imbricated apices, hairs usually short, sparse with *Myzine* type apices but sometime moderately long with blunt, acuminate apices, apterae without secondary rhinaria but alatae always with large, protuberant secondary rhinaria, primary rhinarium ciliated, p.t. slightly to much longer than last antennal segment.
3. Ultimate rostral segment much shorter to much longer than h.t.2. with blunt, obtuse or normal apex, with 2-4 secondary hairs.
4. Midthoracic furca with jointed arms and a short stalk.
5. Dorsum of abdomen rugose, sparsely to strongly spinulose, in apterae pale to sclerotic entirely, without scattered patches, in alatae with dark and central patch, 7th tergite normal or prolonged posteriorly, venter of abdomen strongly spinulose with hairs long and fine.
6. Siphunculus cylindrical to tapering, slight to distinct clavate or at least swollen on distal portion, smooth or strongly imbricated, with or without distinct flange.
7. Cauda may or may not constricted at middle.
8. Subanal plate broadly rounded.
9. Subgenital plates either normal with hind margin rounded or triangular and strongly cauded, with 10-16 short and blunt hairs on posterior margin.
10. Leg, femora sometimes with dense spinules besides normal imbrications, tibiae smooth but highly spinulose, tarsi with normal imbrications, F.T.C. 3,3,2 or 3,3,3.
11. Wing venation normal, veins sometimes bordered brown.

Stroyan (1954), Takahashi (1965c) and Eastop (1966) break up the genus into following subgenera, *Nectaorosiphon, Neotoxoptera, Ptumomyzus, Sciamyzus, Tubaphis* etc. due to variations in characters like the abdominal pigmentation in apterae and alatae, clouded wing veins and presence of secondary rhinaria in alatae only on 3rd segment. However, Miyzaki (1971) considered *Myzus* as a individual genus. From India, eighteen species have been described under this genus.

MYZUS AJARI SP. NOV.

FEMALE: APTERA (Figure 107–112)

2.15 mm long, 0.96 mm broad, greenish yellow; head 0.21 mm long, 0.36 mm broad; antenna 6 segmented, 1.74 mm long, shorter than body; thorax 0.54 mm long, 0.76 mm broad; abdomen 1.34 mm long, 0.96 mm broad; siphunculus 0.34 mm long, 0.06 mm broad, dark greenish yellow; cauda 0.15 mm long, 0.08 mm broad, dark greenish yellow.

HEAD (Figure 107)

0.21 mm long, 0.36 mm broad, greenish yellow, weakly spinulose at both sides; wax pores few, oval, shiny; frons with weakly developed median frontal prominence, lateral frontal tubercles well developed become converging forming shape of 'W'; dorsal cephalic hairs, sparsely hairy with *Myzine* type of apices, hairs shorter than basal diameter of 3rd segment; eyes lateral with distinct triommatidia, ocular space 0.08 mm, interocular space 0.30 mm.

Antenna (Figure 109)

1.74 mm long, pale greenish yellow, scabrous, faintly imbricated, 6 segmented, shorter than body; scape stout, 0.08 mm long, 0.06 mm broad, dark greenish yellow, scabrous, sparsely hairy; pedicel short, 0.04 mm long, 0.04 mm broad, dark greenish yellow, scabrous, sparsely hairy, as long as broad; flagellum 4 segmented, long, distinctly imbricated, sparsely hairy with acuminate apices; 1st segment 0.38 mm long, 0.04 mm broad, without secondary rhinaria; 2nd segment 0.32 mm long, 0.04 mm broad; 3rd segment 0.26 mm long 0.03 mm broad; 4th segment 0.49 mm long, 0.02 mm broad, pale greenish yellow, 4th segment base 0.10 mm long 0.02 mm broad, densely imbricated with serrated margin, primary rhinarium ciliated, round, protuberant at tip base; processus terminalis (p. t.) 0.39 mm long, densely imbricated with serrated margin, p.t. longer than base, 3.9 x base of last antennal segment; antenna shorter than body, 0.80 x body.

Flagellar Formula

1 L/W = 9.5, 2 L/W= 8.0, 4 L/W = 24.0, A = 13.88.

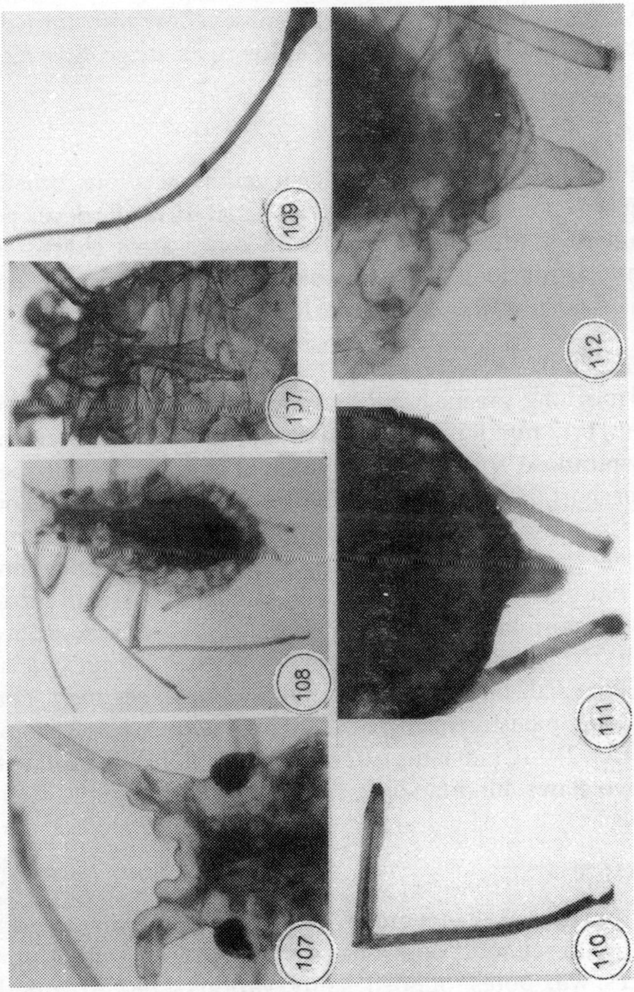

Figures 107-112: *Myzus ajarai* sp.nov. (Female Aptera)
Figure 107: Head and Rostrum; Figure 108: Whole Mount; Figure 109: Antenna; Figure 110: Hind Leg; Figure 111: Siphunculi and Cauda; Figure 112: Cauda

Rostrum (Figure 107)

0.32 long, 0.06 mm broad, greenish yellow, smooth, sparsely hairy, extend upto hindcoxa of metathoracic leg, 5 segmented; 4th and 5th segment fused to form ultimate rostrum segment (u.r.s.), u.r.s. 0.16 mm long, 0.04 mm broad, dark greenish yellow with normal at apex with 4 secondary hairs with fine apices; u.r.s. longer than h.t.2., 1.60 x h.t.2.

THORAX

0.54 mm long, 0.76 broad, greenish yellow, scabrous, broader than long; wax pores rounded, shiny, various sized; faintly developed lateral tubercles, midthoracic furca with joined arms, short stalk; prothorax 0.16 mm long, 0.40 mm broad; mesothorax 0.24 mm long, 0.55 mm broad; metathorax 0.14 mm long, 0.78 mm broad.

Hind Leg (Figure 110)

1.25 mm long, greenish yellow, imbricated, densely spinulose; coxa stout, 0.12 mm long, 0.10 mm broad, dark greenish yellow, densely spinulose, sparsely hairy, broad at base, tapered at apex; trochanter short, 0.04 mm long, 0.06 mm broad, dark greenish yellow, densely spinulose, sparsely hairy; femur 0.29 mm long, 0.06 mm broad, pale green, faintly spinulose with normal imbrications, sparsely hairy; tibia long, narrow, 0.66 mm long, 0.04 mm broad, smooth with normal imbrications, pale green at base, dark green at apex with short hairs at base, long and densely hairy at apex; tarsus 0.10 mm long, 0.02 mm broad, pale green, 1st tarsal segment short, 0.02 mm long, pale greenish yellow, F. T. C. with 3, 3, 2; 2nd tarsal segment (h.t. 2) 0.10 mm long, 0.02 mm broad, slightly curved with curved two claws, hind tarsus 2 (h.t. 2) shorter than u.r.s., 0.60 x u.r.s.

ABDOMEN

1.34 mm long, 0.96 mm broad, dark greenish yellow, sparsely spinulose with sclerotic, dorsal hairs short, sparsely with *Myzine* type apices; wax pores rounded, shiny, various sized, irregularly distributed; 8 segmented; lateral abdominal tubercles weakly developed on 1-7 segment, 1st segment 0.14 mm long, 0.78 mm broad; 2nd segment 0.16 mm long, 0.84 mm broad; 3rd segment 0.16 mm long, 0.86 mm broad; 4th segment 0.18 mm long, 0.96 mm broad; 5th segment 0. 20 mm long, 0.92 mm broad; 6th segment 0.17 mm long,

0.80 mm broad; 7th segment 0.12 mm long, 0.65 mm broad, normal; 8th segment 0.11 mm long, 0.45 mm broad; subanal plate broadly short, convex, rounded; subgenital plate dark, normal, rounded with 14 short, blunt hairs on posterior margin, 4 long fine hairs at anterior margin.

Siphunculus (Figure 111)

0.34 mm long, 0.06 mm broad, dark greenish yellow, cylindrical, strongly imbricated, slightly cleavate, few rows of striae at apex with distinct apical flange; siphunculus longer than cauda, 2.26 x cauda and shorter than body, 0.15 x body.

Cauda (Figure 112)

0.15 mm long, 0.08 mm broad, dark greenish yellow, elongated, slight medially constricted with blunt at apex, faintly spinulose, 6 long hairs with acute apices, cauda shorter than siphunculus, 0.44 x siphunculus.

HOST PLANTS

Tobacco: *Nicotiana tabacum*.

DISCUSSION

Myzus ujuri sp.nov. runs close to *Myzus persicae* Sulzer by having

1. Ultimate rostral segment longer than h.t. 2.
2. Tergite 7 and subgenital plate normal.
3. Dorsal abdominal hairs short with *Myzine* type of apices.
4. Siphunculi with distinct apical flange.

However, differs from above species by following characters

1. Proportion siphunculus to body.
2. Proportion antenna to body.
3. Cauda elongated, blunt at apex with 6 large hairs.
4. Lateral frontal tubercles on head well developed become converging forming shape of 'W', frons with weakly developed median frontal prominence.
5. Ultimate rostral segment longer than h.t. 2.
6. Ultimate rostral segment with normal apex with 4 short hairs.

7. Subgenital plate with 4 long hairs on anterior, 14 short, blunt hairs on posterior margin.
8: Flagellar formula: 1L/W = 9.5, 2 L/W= 8.0, 4 L/W = 24.0, A = 13.88.

CONTROL MEASURES

Natural Control

Ladybird beetles: *Menochilus* sp. cause mortalities in the aphid population.

Chemical Control

Spray: 0.02 per cent phosphamidon or 0.1 per cent carbaryl.

MYZUS TERMINALAE SP. NOV.

FEMALE: APTERA (Figures 113–118)

1.65 mm long, 0.98 mm broad, dark greenish black; head 0.20 mm long, 0.40 mm broad; antenna 6 segmented, 1.10 mm long, shorter than body; thorax 0.54 mm long, 0.90 mm broad; abdomen 0.90 mm long, 0.98 mm broad; siphunculus 0.24 mm long, 0.03 mm broad, dark greenish black; cauda 0.09 mm long, 0.08 mm broad, dark greenish black.

HEAD (Figure 113)

0.20 mm long, 0.40 mm broad, dark greenish black, spinulose at central broad portion and margin dorsally; frons with well developed median frontal prominence become more or less angular moderately developed lateral frontal tubercles with diverging forming shape 'W'; dorsal cephalic hairs short, moderate, with *Myzine* type apices; hairs longer than basal diameter of 3rd segment; eyes lateral with distinct triommatidia, ocular space 0.10 mm, interocular space 0.36 mm.

Antenna (Figure 115)

1.10 mm long, pale green, scabrous, densely imbricated, 6 segmented, shorter than body; scape stout, scabrous, 0.06 mm long, 0.08 mm broad, dark green, sparsely hairy; pedicel short, straight 0.05 mm long, 0.04 mm broad, dark green, scabrous, sparsely hairy; flagellum 4 segmented, densely imbricated, sparsely hairy with *Myzine* type apices; 1st segment 0.26 mm long, 0.03 mm broad, without

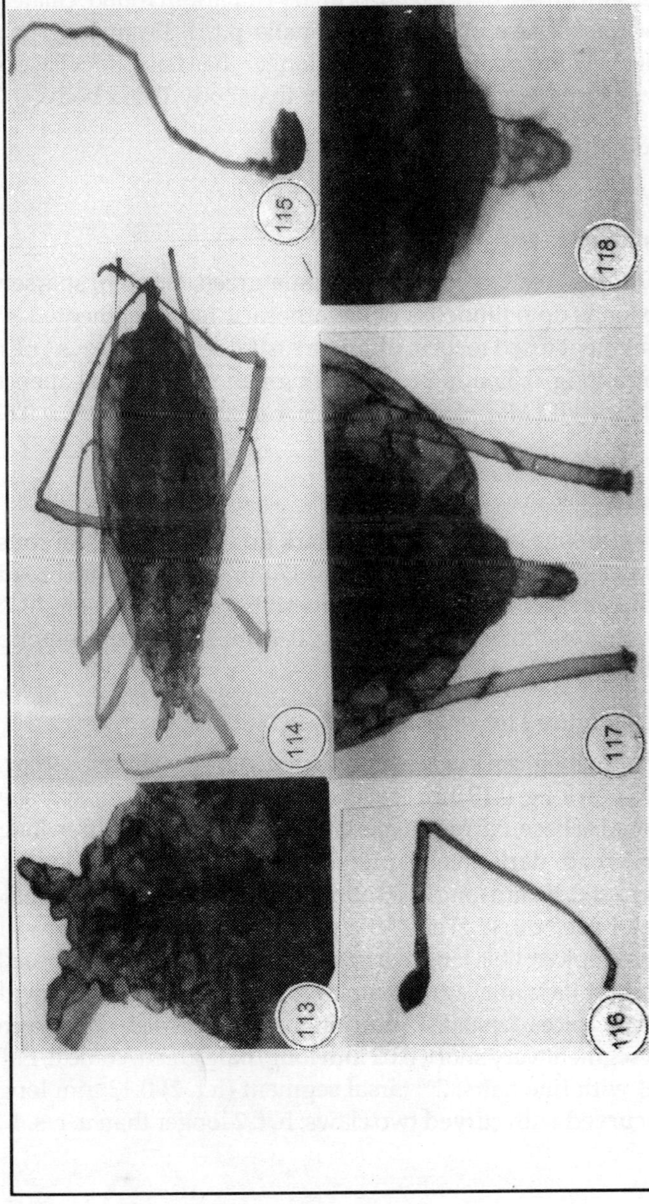

Figures 113-118: *Myzus terminalae* sp.nov. (Female Aptera)
Figure 113: Head and Rostrum; Figure 114: Whole Mount; Figure 115: Antenna;
Figure 116: Hind Leg; Figure 117: Siphunculi; Figure 118: Cauda

secondary rhinaria; 2nd segment 0.17 mm long, 0.03 mm broad; 3rd segment 0.16 mm long, 0.02 mm broad; 4th segment 0.43 mm long, 0.02 mm broad, pale green, imbricated, 4th segment base 0.10 mm long, 0.02 mm broad, dark green, primary rhinarium round, ciliated, placed at apex of base; processus terminalis (p.t.) 0.33 mm long, pale green, densely imbricated, p. t. much longer than base, 3.3 x base of last antennal segment, antenna shorter than body, 0.66 x body.

Flagellar Formula

1 L/W = 8.9, 2 L/W = 5.6, 4 L/W = 24.0, A=12.02.

Rostrum (Figure 113)

0.50 mm long, 0.06 mm broad, pale green, smooth, sparsely hairy, extends upto hindcoxa of metathoracic leg; 5 segmented, 4th and 5th segment fused to form ultimate rostral segment (u. r. s.), u. r. s. 0.10 mm long, 0.05 mm broad, dark green with normal apex, 6 minute hairs with *Myzine* type apices, u. r. s. shorter than h. t. 2, 0.83 x h. t. 2.

THORAX

0.54 mm long, 0.90 mm broad, dark greenish black, scabrous, sparsely hairy, hairs short; prothorax 0.17 mm long, 0.48 mm broad, separate from head and mesothorax; mesothorax 0.21 mm long, 0.78 mm broad; metathorax 0.18 mm long, 0.90 mm broad; mesothorax and metathorax fused.

Hind Leg (Figure 116)

1.48 mm long, dark green, imbricated, densely spinulose; coxa stout 0.08 mm long, 0.10 mm broad, dark green, spinulose, sparsely hairy, broad at base, narrow at apex; trochanter short, 0.06 mm long, 0.06 mm broad, dark green, spinulose, sparsely hairy, as long as broad; femur 0.39 mm long, 0.06 mm broad, faintly imbricated, pale green at base, dark green at apex, long, short hairs with *Myzine* type apices; tibia long, narrow, 0.84 mm long, 0.0.2 mm broad, pale green at base, dark green at apex, long and short hairs with *Myzine* type apices; tarsus 0.12 mm long, 0.02 mm broad, dark green, 1st tarsal segment very short, 0.02 mm long, pale green, conical, F. T. C. 3, 3, 3 with fine hairs; 2nd tarsal segment (h.t. 2) 0.12 mm long, slightly curved with curved two claws; h. t. 2 longer than u. r. s, 1.2 x u. r. s.

ABDOMEN

0.90 mm long, 0.98 mm broad, dark greenish black, highly sclerotic, oval, dorsally and ventrally strongly spinulose, rugose; dorsal hairs long with fine apices, 8 segmented; lateral abdominal tubercles weakly developed on 1–7 segments; 1st segment 0.10 mm long, 0.88 mm broad; 2nd segment 0.12 mm long, 0.90 mm broad; 3rd segment 0.14 mm long, 0.94 mm broad; 4th segment 0.14 mm long, 0.98 mm broad; 5th segment 0.12 mm long, 0.85 mm broad; 6th segment 0.10 mm long, 0.60 mm broad; 7th segment 0.09 mm long, 0.45 mm broad; 8th segment 0.08 mm long, 0.25 mm broad; subanal plate dark green, rounded, convex; subgenital plate dark green, densely spinulose, normal, oval, strongly cauded, 2 pairs long, fine hairs at anteriorly,11 pairs short hairs at posteriorly.

Siphunculus (Figure 117)

0.24 mm long, 0.03 mm broad, dark greenish black, cylindrical, distinctly cleavate, strongly imbricated with faint apical flange; siphunculus longer than cauda, 2.66 x cauda, siphunculus shorter than body 0.14 x body.

Cauda (Figure 118)

0.09 mm long, 0.08 mm broad, dark greenish black, elongated, slight middle constriction, blunt at apex; dense long, short hairs with acute apices; cauda shorter than siphunculus, 0.37 x siphunculus.

HOST PLANTS

Ain: *Terminalia tomentosa.*

DISCUSSION

This species runs close to *Myzus abtusirostris* David, Narayanan and Rajasingh by having

1. Ultimate rostral segment shorter than h.t. 2.
2. Siphunculi without a distinct flange.
3. Siphunculi dark greenish to black.

However, differs from the above species by having following characters:

1. 1st segment of antenna not as wide as long.

2. Ultimate rostral segment with normal apex.
3. Processus terminalis, 3.3 x base of last antennal segment.
4. Lateral frontal tubercles on head moderate diverging forming shape 'W', frons with well developed median frontal prominence become more or less angular.
5. Proportion of siphunculus to cauda.
6. Proportion of antenna to body (1.10 x 1.65).
7. Flagellar formula: 1 L/W = 8.9, 2 L/W = 5.6, 4 L/W = 24.0, A=12.02.

CONTROL MEASURES

Natural Control

Ladybird beetles: *Illeis* sp. feed on aphid populations.

Chemical Control

Spray: 0.03 per cent DDVP or 0.03 per cent malathion.

MYZUS PERSICAE SULZER

FEMALE: ALATA (Figures 119–122, 123–128)

2.60 mm long, 0.95 mm broad, dark brown with blackish tinge; head 0.25 mm long, 0.49 mm broad; antenna 6-segmented, 2.80 mm long, longer than body; thorax 0.81 mm long, 0.72 mm broad; abdomen 1.30 mm long, 0.95 mm broad, dark brown with blackish patches on dorsum; siphunculus 0.60 mm long, 0.06 mm broad, dark brown; cauda 0.23 mm long, 0.14 mm broad, dark brown.

HEAD (Figure 123)

0.25 mm long, 0.49 mm broad, wider than long, with spinulose at dorsal and ventral side; frons with weakly developed median frontal prominence, lateral frontal tubercles well developed its inner margin diverging, forming 'U' shape, dorsal cephalic hairs very short with *Myzine* type of apices; hairs shorter than basal diameter of 3rd segment; eyes lateral with distinct triommatidia, ocular space 0.14 mm, interocular space 0. 41 mm.

Antenna (Figure 126)

2.80 mm long, scabrous, dark brown; 6 segmented, longer than body; scape stout, 0.14 mm long, 0.11 mm broad, dark brown,

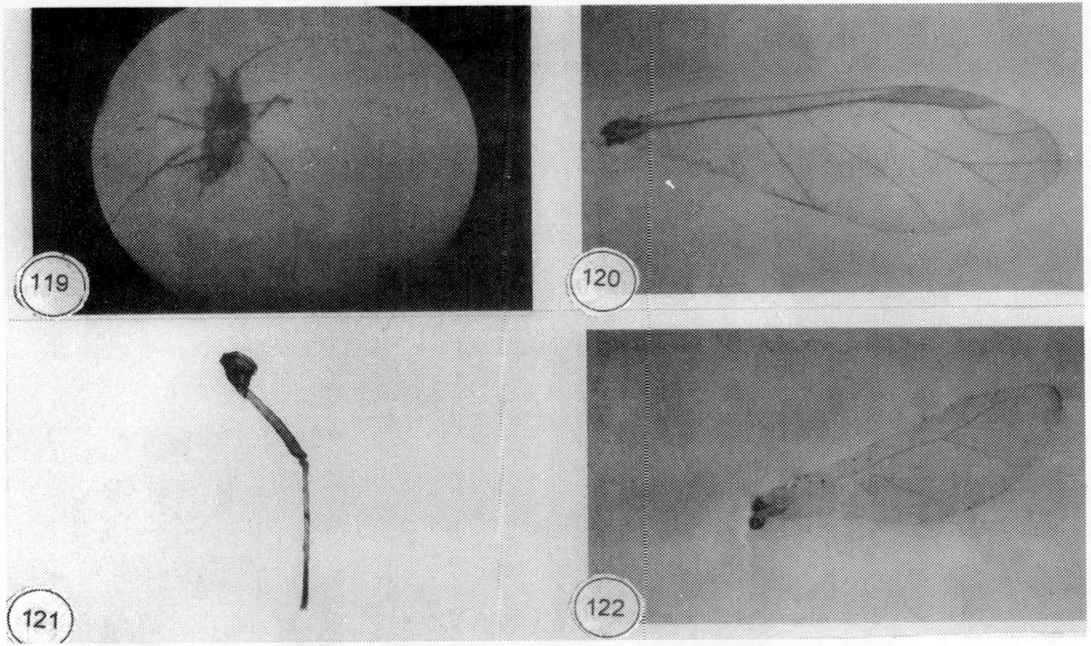

Figures 119-122: *Myzus persicae* sp.nov. (Female Alata)
Figure 119: Female Alata; Figure 120: Fore Wing; Figure 121: Hind Leg; Figure 122: Hind Wing

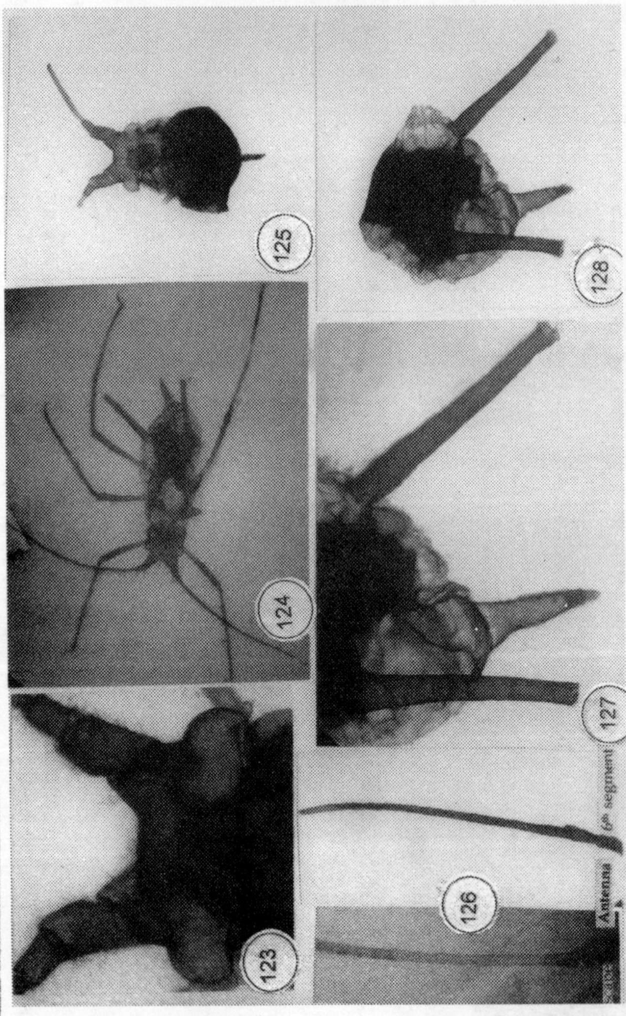

Figures 123-128: *Myzus persicae* sp.nov. (Female Aptera)
Figure 123: Head; Figure 124: Whole Mount; Figure 125: Rostrum; Figure 126: Antenna; Figure 127: Siphunculi; Figure 128: Cauda

scabrous with sparsely hairy; pedicel short, straight, 0.09 mm long, 0.07 mm broad, scabrous, sparsely hairy; flagellum 4 segmented, densely imbricated, sparsely hairy with *Myzine* type apices; 1st segment 0.82 mm long, 0.05 mm broad, densely imbricated with small to large many protuberant secondary rhinaria; 2nd segment 0.47 mm long, 0.04 mm broad; 3rd segment 0.39 mm long, 0.03 mm broad, 4th segment 0.84 mm long, 0.02 mm broad, 4th segment base short, 0.15 mm long, 0.03 mm broad, primary rhinarium ciliated, round at tip of base; processus terminalis (p.t.) 0.70 mm long, dark brown, longer than base, 4.66 x base of last antennal segment; antenna longer than body, 1.06 x body.

Flagellar Formula

1 L/W= 16.4, 2 L/W= 11.75, 4 L/W = 42.0, A= 28.33.

Rostrum (Figure 123, 125)

0.82 mm long, 0.05 mm broad, dark brown, smooth, extends upto hindcoxa of metathoracic leg, sparsely hairy, 5 segmented; 4th and 5th segment fused to form ultimate rostral segment (u.r.s.), 0.20 mm long, 0.04 mm broad, u.r.s. normal with 2 long, 2 short hairs, u.r.s. longer than h.t.2, 1.11 x h.t.2.

THORAX

0.81 mm long, 0.72 mm broad, dark brown with blackish tinge, sparsely hairy with *Myzine* type of apices, midthoracic furca with jointed arms and short stalk; prothorax 0.19 mm long, 0.44 mm broad; mesothorax 0.32 mm long, 0.64 mm broad; metathorax 0.28 mm long, 0.72 mm broad.

Fore Wing (Figure 120)

Fore wing 3.20 mm long, 1.10 mm broad, membranous, transparent, faintly spinulose at periphery; pterostigma, radial sector (Rs), Media (M), Cubitus (Cu) and anal veins (A) dark brown, media bisected, cubitus and anal veins straight.

Hind Wing (Figure 122)

Hind wings membranous, transparent, 1.60 mm long, 045 mm broad, shorter than fore wing; consists two oblique veins cubitus, anal veins, become parallel, straight and reached upto margin of wing.

Hind Leg (Figure 121)

2.34 mm long, pale brown; coxa stout, 0.14 mm long, 0.12 mm broad, dark brown, sparsely hairy, broad at base, narrow at apex; trochanter short, triangular, 0.09 mm long, 0.08 mm broad, dark brown, sparsely hairy; femur 0.66 mm long, 0.06 mm broad, straight, densely spinulose with normal hairs, pale brown at base, dark brown at apex; tibia 1.32 mm long, 0.04 mm broad, densely spinulose with primary and secondary hairs with *Myzine* apices; tarsus 0.18 mm long, 0.04 mm broad, pale brown with normal imbrications, sparsely hairy, 1st tarsal segment short, 0.03 mm long, pale brown, conical, F. T. C. with 3,3,3; 2nd tarsal segment (h.t.2) 0.18 mm long, slightly curved, dark brown with curved two claws; h.t.2 shorter than u.r.s., 0.90 x u.r.s.

ABDOMEN

1.30 mm long, 0.95 mm broad, dark brown, rugose, spinulose, dark large central patch beside marginal sclerotic patches, dorsal abdominal hairs short, sparsely with *Myzine* type apices; 8 segmented; 1st segment 0.14 mm long, 0.62 mm broad; 2nd segment 0.16 mm long, 0.70 mm broad; 3rd segment 0.17 mm long, 0.82 mm broad; 4th segment 0.20 mm long, 0.90 mm broad; 5th segment 0.19 mm long, 0.95 mm broad; 6th segment 0.15 mm long, 0.81 mm broad; 7th segment 0.14 mm long, 0.70 mm broad, 7th tergite prolonged posteriorly; 8th segment 0.14 mm long, 0.55 mm broad, 8th tergite possess uniform median hairs; distance between 6th and 7th spiracles distinctly longer than that between 5th and 6th; Subanal plate broadly rounded; subgenital plate normal, rounded, hind margin with 10 short hairs.

Siphunculus (Figure 127)

0.60 mm long, 0.06 mm broad, cylindrical, smooth, without hairs, strongly clavate, distinct apical flange; siphunculus longer than cauda, 2.60 x cauda; shorter than body, 0.23 x body.

Cauda (Figure 128)

Cauda 0.23 mm long, 0.14 mm broad, elongated, broad at base, globular medially and tapered at apex, without middle constriction with 4 primary, 4 secondary hairs with *Myzine* type apices; cauda shorter than siphunculus, 0.38 x siphunculus.

Figure 129: ***Myzus ajari*** **sp.nov.; Figure 130:** ***Myzus terminalae*** **sp.nov.; Figure 131:** ***Myzus terminalae*** **sp.nov.; Figure 132:** ***Myzus persicae***

HOST PLANTS

Brinjal–*Solanum melongena*, Cabbage–*Brassica capitata;* Carrot–*Dancus carrota;* Bryophylum–*Bryophylum pinnatum;* Khair–*Acacia catechu;* Teak–*Tectona grandis.*

CONTROL MEASURES

Natural Control

Tiger beetles, *Eudema angulatum, Menochilus* sp., lace wings feeds on pest populations.

Chemical Control

0.03 per cent DDVP spray or 0.03 per cent Azadirachtin spray or 0.03 per cent malathion spray.

GENUS *MACROSIPHUM* PASSERINI

The genus was erected by Passerini in 1860. He used *Aphis rosae* Linnaeus as a type species. It is cosmopolitan in distribution and characterized by:

1. Head completely smooth, minute spinulose either on both surfaces or on one surface with moderate or strongly developed.
2. Lateral frontal tubercles smooth, diverging, with indistinct or distinct frontal prominence.
3. Dorsal cephalic hairs short or long, with acute or blunt apices.
4. Antenna 6 segmented, shorter or longer than the body, basal 2 segments either smooth or scabrous, flagellum is imbricated, apterae may secondary rhinaria restricted base of 3rd segment or distributed entire length of segment, alatae such rhinaria present on 3rd segment, also on 4th segment, primary rhinaria ciliated, process terminalis (p.t.) much longer than base of last antennal segment.
5. Ultimate rostral segment shorter or longer than h.t.2., u.r.s. heart shaped or elongate with blunt apex with 4-6 secondary hairs.
6. Leg smooth, hind tibiae with spines besides normal hairs, second tarsal segments with normal imbrications.

7. Wing venation normal with stigma, veins sometimes dark.
8. Abdominal dorsum in apterae completely pale or sometime pigmented middorsally and marginally upto 6th segment, in alatae, pale with brown marginal patches and transverse pleural bands linked by sclerotic area, venter of abdomen with spinulose striae, ventral abdominal hairs long with fine apices.
9. Siphunculus cylindrical, with or without distinct flange, imbrications passing onto distinct apical reticulated area, imbrications faint or distinct.
10. Cauda elongate, pale or dark, with or without a basal constriction.
11. Subanal plate is somewhat emarginate on the posterior margin.
12. Subgenital plate with 7-16 hairs in 2 groups on posterior margin and 2 long fine hairs on anterior.

Lambers (1939) discussed the genus *Macrosiphum* in detail. He didn't mention spinulosity on head and F.T.C. Takahashi (1961b) erected genus *Unisitobion* for *Macrosiphum* like species possessing spinulosity on the venter of the head, pigmented abdominal dorsum and high antennal sockets but such species have F.T.C. 3,3,3; since all *Macrosiphum* like species except the Indian species, namely *Macrospinulosum* David, Rajasingh and Narayanan, *Pseudoluteum* Ghosh, and *Pseudogeranii* Chakrabarti and Raychaudhuri have F.T.C. 4,4,4; a new subgenus *Neomacrosiphum* is recognized by Raychaudhuri, 1980 to accommodate them with *Macrosiphum pseudoluteum* Ghosh as the type.

According to Raychaudhuri (1980), *Unisitobion* is considered as synonym of *Macrosiphum*. Lambers (1939) has mentioned high lateral frontal tubercles for *Macrosiphum* species, then only difference between typical *Macrosiphum* and *Unisitobion* lies in absence and presence of spinules on head. Takahashi (1961) didn't consider *rubiphilla* under his new genus. If *Unisitobion* Takahashi is considered a distinct genus, one should expect to get a combination of characters as mentioned by Takahashi (1961). Indian specimens having spinules on head, available now do not show dark pigmented body. Moreover, if one looks at *Myrus obtusirostris* David, Naryananan and Rajasingh, the idea of body pigmentation in *Myrus* is altogether changed. From India 17 species have been described under this genus.

MACROSIPHUM KOYENENSIS SP. NOV.

FEMALE: ALATA (Figures 133–141)

2.60 mm long, 0.98 mm broad, brown, smooth; head 0.22 mm long, 0.44 mm broad; thorax 0.80 mm long, 0.60 mm broad; abdomen 1.45 mm long, 0.98 mm broad; siphunculus 0.59 mm long, 0.06 mm broad, dark brown; cauda 0.15 mm long, 0.06 mm broad, pale brown.

HEAD (Figure 137)

0.22 mm long, 0.44 mm broad, pale brown, smooth, lateral frontal tubercles indistinct, diverging, frons with distinct median frontal prominence forming weakly shape of 'W', 4 dorsal cephalic hairs sparsely hairy, hairs short with acute apices; eyes lateral with distinct triommatidia, ocular space 0.12 mm, interocular space 0.36 mm.

Antenna (Figure 139)

2.35 mm long, pale brown, smooth, sparsely hairy, 6 segmented, shorter than body; scape stout 0.10 mm long, 0.06 mm broad, slight dark brown, smooth, sparsely hairy; pedicel 0.08 mm long, 0.06 mm broad, slight dark brown, smooth, sparsely hairy; flagellum long, 4 segmented, faintly imbricated, pale brown; 1st segment 0.66 mm long, 0.06 mm broad, secondary rhinaria in row restricted at very base; 2nd segment 0.48 mm long, 0.04 mm broad, secondary rhinaria absent; 3rd segment 0.40 mm long, 0.04 mm broad with 4 secondary rhinaria present; 4th segment 0.71 mm long, 0.03 mm broad, dark brown, faintly imbricated, 4th segment base short, 0.14 mm long, 0.03 mm broad, primary rhinarium ciliated, round at tip of base; processus terminalis (p.t.) 0.57 mm long, 0.02 mm broad, dark brown, imbricated with serrated margin, without hairs; p.t. longer than base, 4.14 x base of last antennal segment; antenna shorter than body, 0.90 x body.

Flagellar Formula

1 L/W = 11.0, 2 L/W = 12.0, 4 L/W = 23.66, A = 15.53.

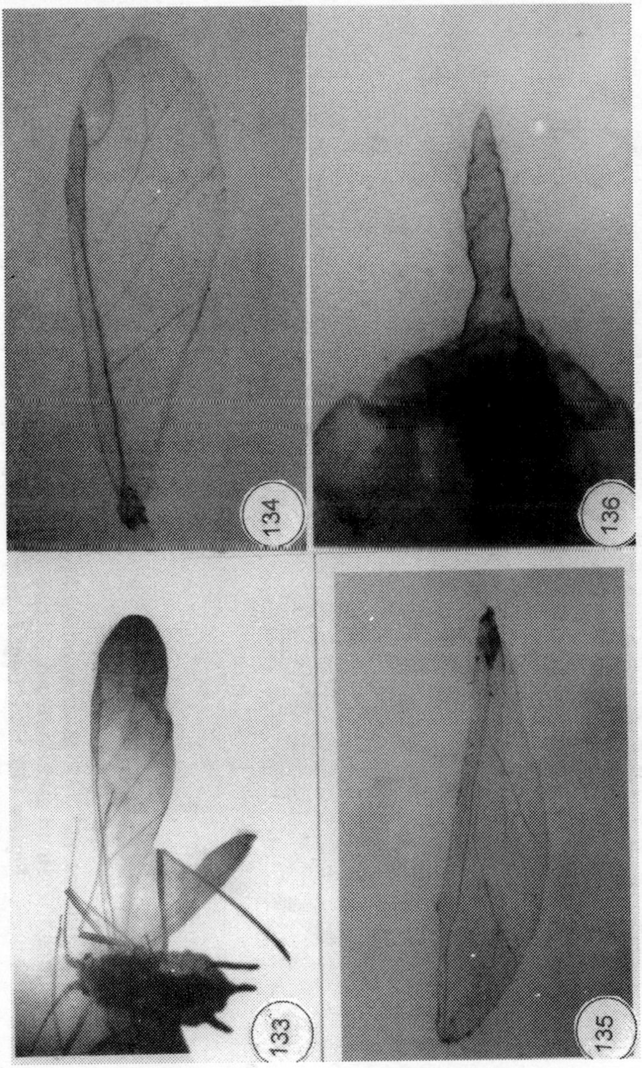

Figures 133-136: ***Macrosiphum koyenensis*** **sp.nov. (Female Alata)**
Figure 133: Female Alata; Figure 134: Fore Wing; Figure 135: Hind Wing; Figure 136: Cauda

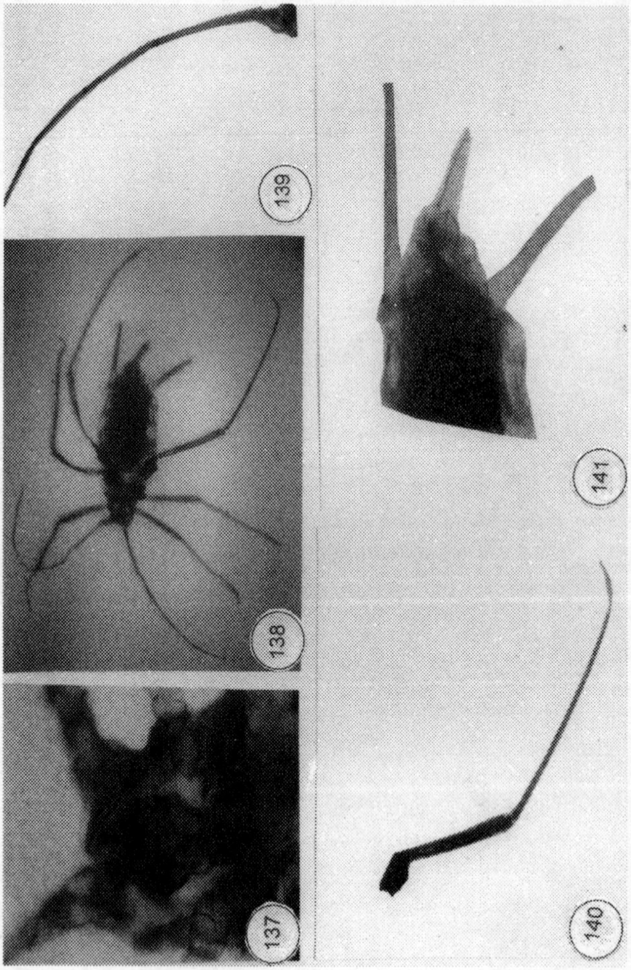

Figures 137-141: *Macrosiphum koyenensis* sp.nov. (Female Aptera)
Figure 137: Head and Rostrum; Figure 138: Whole Mount; Figure 139: Antenna; Figure 140: Hind Leg; Figure 141: Siphunculi and Cauda

Rostrum (Figure 137)

0.60 mm long, 0.04 mm broad, dark brown, smooth, extend upto hindcoxa of metathoracic leg, 5 segmented; 4th and 5th segment fused to form ultimate rostral segment (u.r.s.), u.r.s. 0.12 mm long, 0.04 mm broad, dark brown, elongated, bluntish, with 3 primary, 6 secondary hairs with acute apices; u.r.s. shorter than h.t.2, 0.85 x h.t.2.

THORAX

0.80 mm long, 0.60 mm broad, pale brown, smooth, longer than broad; prothorax 0.20 mm long, 0.41 mm broad, with dorsal sclerotic plate smooth, without hairs, separate from head, mesothorax; mesothorax well developed, 0.30 mm long, 0.60 mm broad, continuous with complete sclerotic plate, sclerotic plate pale brown, smooth, globular, broad anteriorly narrow posteriorly; metathorax 0.30 mm long, 0.60 mm broad, sclerotic plate of metathorax pale brown, smooth, slight sculptured, without hairs.

Hind Leg (Figure 140)

2.90 mm long, pale brown, smooth; coxa stout, 0.20 mm long, 0.11 mm broad, smooth; trochanter short, 0.10 mm long, 0.09 mm broad, smooth; femur 0.80 mm long, 0.08 mm broad, straight, smooth, without hairs, pale brown at base, dark brown at apex; tibia 1.58 mm long, 0.05 mm broad, pale brown, smooth, straight, pale brown at base, dark brown at apex, densely hairy; tarsus 0.14 mm long, 0.04 mm broad 1st tarsal segment short, 0.02 mm long, conical, pale brown, F.T.C. with 3,3,3 hairs; 2nd tarsus 0.14 mm long, 0.03 mm broad, with normal imbrications, slightly curved with curved two claws; h.t.2 longer than u.r.s., 1.16 x u.r.s.

Fore Wing (Figure 134)

2.90 mm long, 0.91 mm broad, pale, transparent, faint spinulose at periphery of wing, pterostigma dark, thick, brown; radial sector (Rs) curved touch to periphery; media straight, biforked reached just near margin, not touching to pterostigma anteriorly; cubitus (Cu) straight arises from pterostigma upto margin; anal vein (A) straight start from pterostigma upto margin. All wing venations normal, dark brown.

Hind Wing (Figure 135)

0.70 mm long, 0.22 mm broad, pale, transparent, spinulose at margin, two oblique veins cubitus (Cu) and Anal vein (A) straight, run from pterostigma, reach very close to margin of wing, oblique veins slight dark brown.

ABDOMEN

1.45 mm long, 0.98 mm broad, pale brown, smooth with central dark patch extend upto siphunculus, brown marginal patches with transverse pleural bands, sparsely hairy, venter with spinulose striae, hairs long with fine apices, 8 segmented; lateral abdominal tubercles on 1-7 segments, spiracles black situated laterally; 1st segment 0.16 mm long, 0.76 mm broad; 2nd segment 0.16 mm long, 0.80 mm broad; 3rd segment 0.18 mm long, 0.86 mm broad; 4th segment 0.20 mm long, 0.96 mm broad; 5th segment 0.16 mm long, 0.98 mm broad; 6th segment 0.12 mm long, 0.88 mm broad; 7th segment 0.14 mm long, 0.81 mm broad; 8th segment 0.16 mm long, 0.71 mm broad, dense, stiff, long hairs; subanal plate emarginate on posterior margin; subgenital plate with 14 hairs in 2 groups on posterior margin.

Siphunculus (Figure 141)

0.59 mm long, 0.06 mm broad, cylindrical, dark brown, smooth, without apical flange at tip, faintly imbricated, imbrications passing into apical distinct reticulated area, without hairs; siphunculus longer than cauda, 4.21 x cauda; shorter than body, 0.22 x body.

Cauda (Figure 136, 141)

0.15 mm long, 0.06 mm broad, pale brown, elongated, conical, with basally constricted, 6 long, primary hairs, many secondary hairs; cauda shorter than siphunculus, 0.23 x siphunculus.

HOST PLANTS

Rose: *Rose multiform* Thumb.

DISCUSSION

This species runs close to *Macrosiphum spinotibium* Ghosh, Ghosh and Raychaudhuri (Raychaudhuri, 1980) by having following characters:

1. First tarsal segment with 3, 3, 3 hairs.
2. Hind tibia with short spine like hairs besides normal hairs.

However, it differs from the above species by having following characters:

1. Hind tibia is without row of short spine like hairs besides normal hairs.
2. Ultimate rostral segment without 4 secondary hairs.
3. Body length is shorter, 1.28 mm.
4. Siphunculus 0.59 mm long, straight, cylindrical.
5. Cauda 0.14 mm long, acute, elongated.
6. Lateral frontal tubercles on head indistinct, diverging, frons with distinct median frontal prominence forming weak shape of 'W'.
7. 4 dorsal cephalic hairs short with acute apices.
8. Flagellar formula: 1 L/W = 11.0, 2 L/W = 12.0, 4 L/W = 23.66, A = 15.53.

CONTROL MEASURES

Natural Control

Ladybird beetles: *Microphis* sp., *Coccinella* sp., lacewing feeds on aphids.

Chemical Control

Spray: 0.02 per cent phosphamidon (or) 0.03 per cent malathion (or) 0.05 per cent Azadirachtin.

GENUS *NEOMYZUS* VAN DER GOOT

Van der Goot (1915) erected this genus with the help of a type species *Siphonophora circumflex* Buckton. The genus is distributed in America, Australia, Europe, India, Japan, Java and New Zealand. It is characterized by having following features:

1. Head strongly spinulose, with slightly to well developed laterally frontal tubercles, without median frontal prominence but sometimes slight convex.
2. Antenna 6 segmented, shorter or longer than body, 1st and 2nd segment scabrous, weakly warty, outer margin of 3rd segment smooth while inner margin moderately to distinctly imbricated; hairs on 3rd segment very short with blunt to incrassate apices; in apterae without or with 3

secondary rhinaria, in alatae with many secondary rhinaria on 3rd and 4th segment, p.t. 3.20- 5.0 x base of last antennal segment.

3. Ultimate rostral segment normal in shape, shorter to longer than h.t. 2 with 2-6 secondary hairs.
4. In apterae, dorsum of abdomen smooth or rugose, pale or pigmented, in alatae with solid dorsal abdominal black patch over tergites 3-6 and transverse pigmented on 1st segment, dorsal hairs short and inconspicuous with apices like those on 3rd segment.
5. Siphunculus cylindrical or clavate distally, 0.15–0.25 x body, longer than cauda with 4-7 hairs.
6. Subgenital plate broadly oval, with 8–10 hairs on posterior margin and with 4 hairs at anterior margin.
7. Legs light yellowish to brown but always pale basally in two segments and basal 0.50 portions of femur, femur smooth, sometimes with spinulose striae, tibiae smooth, F.T.C. 3, 3, 3, or 4, 4, 4.
8. Wing venation normal.

According to Cottier (1953) the genus *Neomyzus* was considered close to *Aulacorthum*, but Hille Ris Lambers (1947) placed *Neomyzus* under *Aulacorthum* as a subgenus. Borner (1952), Eastop (1958, 1966) and Heinze (1961) gave generic status to *Neomyzus* due to presence of horse shoe shaped black dorsal patch on dorsum. In India, four species have been described under this genus.

NEOMYZUS KOYANI SP. NOV.

FEMALE: APTERA (Figures 142–147)

1.90 mm long, 0.89 mm broad, greenish black; head 0.20 mm long, 0.34 mm broad; antenna 6 segmented, 1.10 mm long, shorter than body; thorax 0.60 mm long, 0.72 mm broad; abdomen 1.20 mm long, 0.89 mm broad; siphunculus 0.34 mm long, 0.06 mm broad, pale greenish black; cauda 0.10 mm long, 0.06 mm broad, pale greenish black.

HEAD (Figure 142)

0.20 mm long, 0.34 mm broad, dark greenish black, densely spinulose, dorsoventrally flat; lateral frontal tubercles weakly

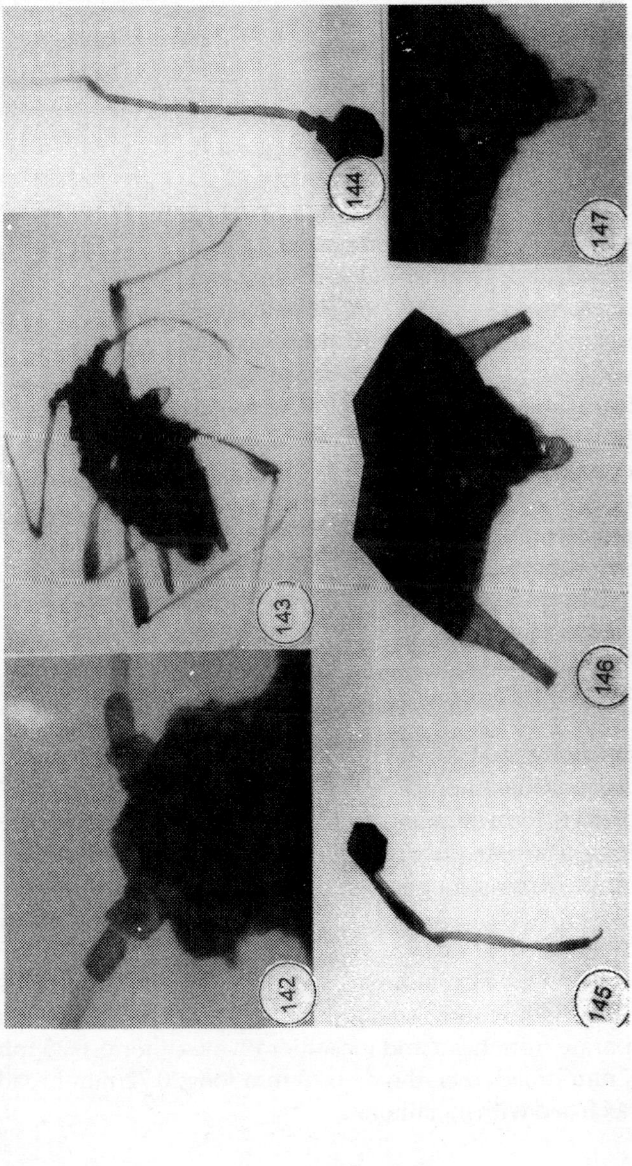

Figures 142-147: *Neomyzus koyani* sp.nov. (Female Aptera)
Figure 142: Head and Rostrum; Figure 143: Whole Mount; Figure 144: Antenna; Figure 145: Hind Leg; Figure 146: Siphunculi; Figure 147: Cauda

developed become diverging type, frons with well developed median frontal prominence forming triangular shape, dorsal cephalic hairs short, sparsely with acute apices; eyes lateral with indistinct triommatidia,ocular space 0.08 mm, interocular space 0.30 mm.

Antenna (Figure 144)

1.10 mm long, pale greenish black, scabrous, densely imbricated, 6 segmented, shorter than body; scape stout, 0.06 mm long, 0.08 mm broad, dark greenish black, scabrous, sparsely hairy with acute apices; pedicel 0.04 mm long, 0.06 mm broad, dark greenish black, scabrous sparsely hairy with acute apices; flagellum 4 segmented, pale greenish black, distinctly imbricated; 1st segment 0.26 mm long, 0.03 mm broad, dark greenish black, distinctly imbricated, sparsely hairy with incrassate apices, longest hair on 3rd segment shorter than 0.55x basal diameters of 3rd segment, secondary rhinaria absent; 2nd segment 0.19 mm long, 0.03 mm broad; 3rd segment 0.18 mm long, 0.02 mm broad; 4th segment 0.39 mm long, 0.02 mm broad, pale green, densely imbricated, 4th segment base short, 0.10 mm long, 0.02 broad, primary rhinarium ciliated, round, protuberant at tip of base; processus terminalis (p.t.) long, 0.29 mm long, pale greenish black, distinctly imbricated; p.t. longer than base, 2.9 x base of last antennal segment; antenna shorter than body, 0.57 x body.

Flagellar Formula

1 L/W =8.6, 2 L/W = 6.3, 4 L/W = 19.00, A = 11.21.

Rostrum (Figure 142)

0.40 mm long, 0.06 mm broad, smooth, sparsely hairy, extend upto hindcoxa of metathoracic leg, rostrum 5 segmented; 4th and 5th segment fused to form ultimate rostral segment (u.r.s.), u.r.s. normal, 0.14 mm long, 0.06 mm broad, dark blackish with 6 secondary hairs with acute apices, u.r.s. longer than h.t.2, 1.4 x h.t.2.

THORAX

0.60 mm long, 0.72 mm broad, dark greenish black, sclerotic, sparsely hairy with acute apices; prothorax 0.17 mm long, 0.42 mm broad, separate from head and mesothorax; mesothorax 0.25 mm long, 0.56 mm broad; metathorax 0.20 mm long, 0.72 mm broad, mesothorax fused with metathorax.

Hind Leg (Figure 145)

1.26 mm long, pale greenish black, faintly imbricated; coxa stout, 0.08 mm long, 0.10 mm broad, pale green with sparsely hairy, broad at base, narrow at apex; trochanter short, 0.06 mm long, 0.06 mm broad, pale green with sparsely hairy; femur long, 0.39 mm long, 0.06 mm broad, faintly scaly, imbricated, slight curved, pale green at base, dark green, broad at apex, sparsely hairy with acute apices; tibia long, narrow, 0.69 mm long, 0.04 mm broad, pale green at base, dark green at apex, densely hairy with long and short hairs with acute apices; tarsus short 0.10 mm long, smooth with normal imbrications, 1st tarsal segment, 0.02 mm long, smooth, pale green, F.T.C. with 3,3,3; 2nd tarsal segment 0.10 mm long, slightly curved with curved two claws, h.t.2 with 4 dorsolateral short hairs with acute apices, h.t.2 shorter than u.r.s., 0.24 x u.r.s.

ABDOMEN

1.20 mm long, 0.89 mm broad, greenish black, faintly rugose with few pigmentations, pigmented area on antisiphuncular segment forms horse shoe shaped patches, dorsal hairs short with incrassate apices, longest hair on anterior tergites longer than basal diameter of 3rd segment, 8 segmented; lateral abdominal tubercles weakly developed on 1 -7 segments, 1st segment 0.12 mm long, 0.80 mm broad; 2nd segment 0.14 mm long, 0.85 broad; 3rd segment 0.18 mm long, 0.86 mm broad; 4th segment 0.19 mm long, 0.89 mm broad; 5th segment 0.13 mm long, 0.86 mm broad; 6th segment 0.12 mm long, 0.80 mm broad; 7th segment 0.10 mm long, 0.50 mm broad, 8th segment 0.08 mm long, 0.24 mm broad; subanal plate globular, semioval, dark green; subgenital plate broadly oval, dark green with 12 pairs of hairs at posterior margin, 5 pairs of hairs at anterior margin with incrassate apices.

Siphunculus (Figure 146)

0.34 mm long, 0.06 mm broad, pale greenish black, cylindrical, faintly imbricated, broad at base, slight tapered at apex; siphunculus longer than cauda, 3.4 x cauda and shorter than body, 0.18 x body.

Cauda (Figure 147)

0.10 mm long, 0.06 mm broad, pale greenish black, faintly imbricated with blunt apex, densely hairy long and short hairs with acute apices; cauda shorter than siphunculus, 0.27 x siphunculus.

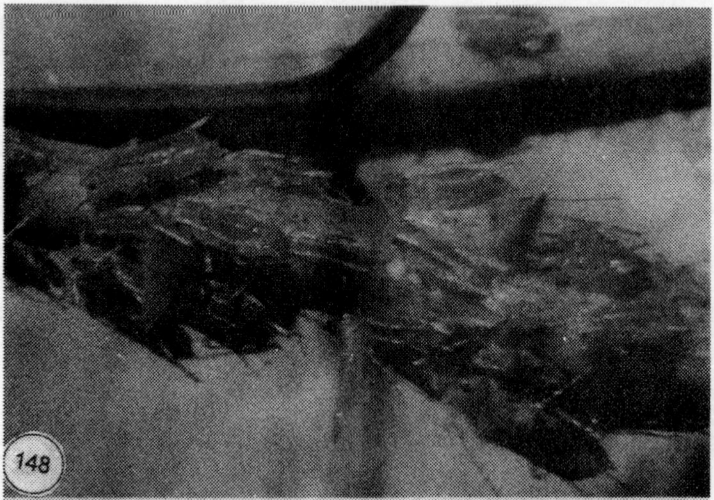

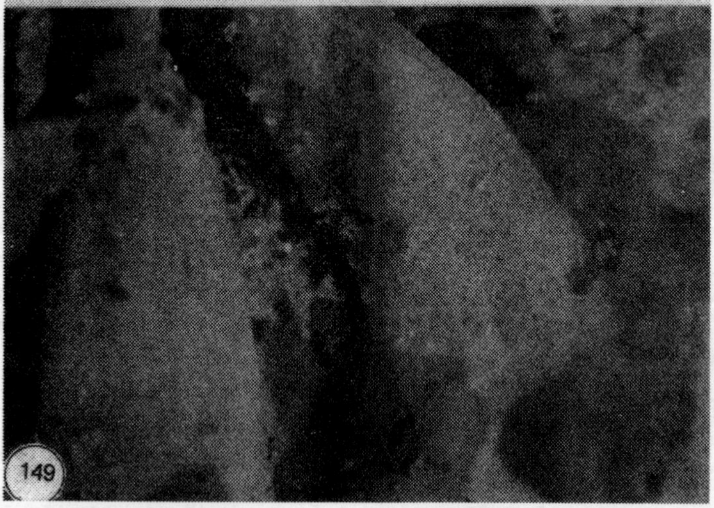

Figures 148: Colony of *Macrosiphum koyenensis* sp.nov.
Figure 149: Colony of *Neomyzus koyani* sp.nov.

HOST PLANTS

Piwali Putani: *Ticoma stans*.

DISCUSSION

Neomyzus koyani runs close to *Neomyzus circumflexus* Buckton by having:

1. Siphunculi usually cylindrical.
2. Fore femora with scaly imbrications on surface.
3. Flagellum with prominent scaly imbrications.

However, it differs from above species by following characters -

1. Siphunculi broad at base, tapered at apex.
2. Antenna, siphunculi and cauda pale.
3. Lateral frontal tubercles on head weakly developed, become diverging, frons with well developed median prominence forming triangular shape.
4. Proportion of siphunculus to cauda.
5. Antenna shorter than body.
6. Proportion of siphunculus to body.
7. Subgenital plate oval with 5 pairs of hairs at anterior margin and 12 pairs of hairs at posterior margin.
8. h.t. 2 with 4 dorsolateral short hairs with acute apices.
9. Flagellar formula: 1 L/W =8.6, 2 L/W = 6.3, 4 L/W = 19.00, A = 11.21.

CONTROL MEASURES

Natural Control

Ladybird beetles: *Menochilus* sp., *Vernia* sp. feed on aphids.

Chemical Control

Spray: 0.03 per cent phosphamidon (or) 0.03 per cent malathion (or) 0.03 per cent Azadirachtin.

GENUS *EUTRICHOSIPHUM* ESSIA AND KUWANA

The genus was erected by Essig and Kuwana in 1918 on the basis of type species *Trichosiphum pasoniae* Okajima. The genus is distributed in China, India, Indonesia, Japan, Malaysia, Nepal, Pakistan, Taiwan and Thialand. It is characterised by following features.

1. Body is elongated, oval in shape.
2. Head, frons straight without lateral frontal tubercles.
3. Antenna 5 or 6 segmented, shorter than body, basal segment, scape and pedicel smooth, flagellum gradually more distinctly imbricated, long or short hairs on flagellum with acute, acuminate apices, alatae secondary rhinaria usually on 3rd segment and sometimes on 4th segment, p.t. as long as to much longer than base of last antennal segment
4. Rostrum long with distinct 4th and 5th segments, u.r.s. about 1.50–3.60 x h.t.2.
5. Legs pale, femur ventrally with spinulose striae, tibia mostly smooth except imbricated stripes, 1st tarsal segment with 7 hairs.
6. Wing venation normal.
7. Dorsum of abdomen in apterae may smooth entirely or medially or entirely spinulose; hairs on tergites long and short, stout and stiff or fine with acute or furcated apices.
8. Venter of abdomen with spinulose striae.
9. Siphunculus elongated, outwardly curved, constricted at base, non reticulated in apterae but faintly striated in alatae, hairs mostly uniform length with fine apices but few hairs with furcated apices, apterae spines present entire length but more dense towards apex of siphunculus.
10. Cauda broadly oval.
11. Subanal plate round.

In India 21 species have been described under this genus.

Paratichosiphum alnicola Basu, A.N., Bull. Ent. (1967), 18: 14.

Eutrichosiphum (*Paratrichosiphum*) *alnicola,* Raychaudhuri D.N. and Chaterjee M. In "Taxonomy of aphids of North India and Bhutan" (ed.) Raychaudhuri, D.N. (1980), 356.

Eutrichosiphum (*Parathrichosiphum*) *alnicola,* Raychaudhuri D.N., Ghosh, L.K. and Das S.K., Ins. matsum; n.s. (1980), 20: 64.

Eutrichosiphum (*Paratrichosiphum*) *alnicola*, Saha, S. and Chakrabarthi, S., Proc. Zool. Soc. Calcutta (1984), 37: 48.

This genus is subdivided into three subgenera as follows:

1. *Eutrichosiphum*
2. *Paratrichosiphum*
3. *Neoparatrichosiphum*

The subgenus *Eutrichosiphum* is characterised by following features:

1. Antenna 5 segmented.

The subgenus *Paratrichosiphum* is recognized by the following features:

1. Antenna 6 segmented.
2. Hairs on siphunculi fine and more or less of similar length.

The subgenus *Neoparatrichosiphum* is visualized by following features:

1. Antenna 6 segmented.
2. Some hairs on siphunculus with fine apices, some with furcated apices, later may extend upto basal 0.70 portion of siphunculus.

From India 21 species of *Eutrichosiphum* have been described.

EUTRICHOSIPHUM PARATRICHOSIPHUM PISIDIUMI SP.NOV.

FEMALE: APTERA (Figures 150–155)

2.14 mm long,1.05 mm broad, dark brown; head 0.29 mm long, 0.48 mm broad; antenna 6 segmented, 1.60 mm long, shorter than body; thorax 0.59 mm long, 0.92 mm broad; abdomen 1.20 mm long, 0.92 mm broad, dark brown; siphunculus 0.60 mm long, 0.10 mm broad, dark brown; cauda 0.12 mm long, 0.24 mm broad, pale brown.

HEAD (Figure 153)

0.29 mm long, 0.48 mm broad, dark brown, smooth, many long hairs with acuminate apices; frons with slightly median frontal

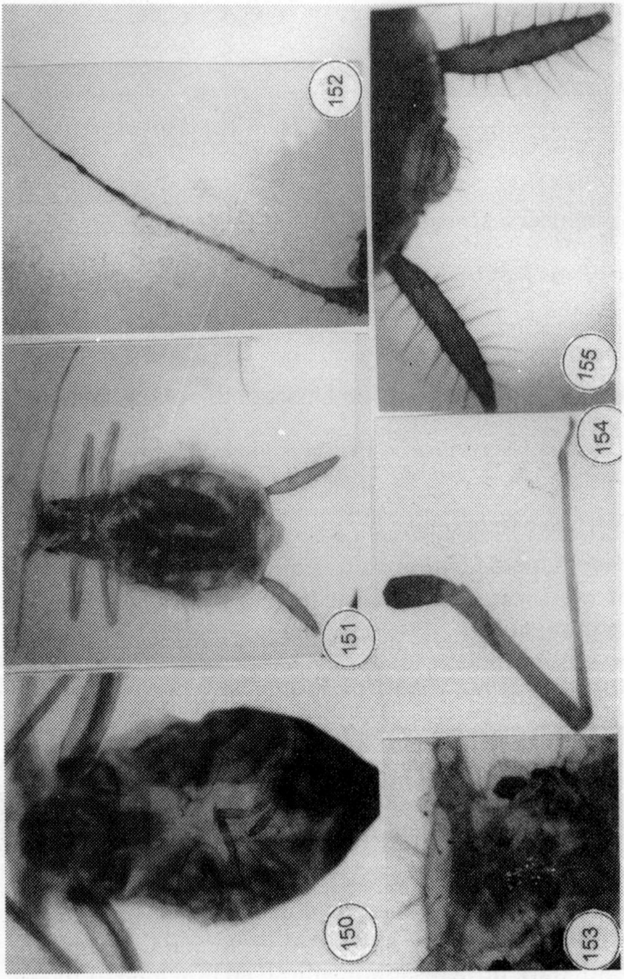

Figures 150-155: *Eutrichosiphum paratrichosiphum pisidiumi* sp.nov. (Female Aptera)
Figure 150: Rostrum; Figure 151: Whole Mount; Figure 152: Antenna; Figure 153: Head; Figure 154: Hind Leg; Figure 155: Siphunculi and Cauda

prominence, without lateral frontal tubercles, eyes lateral with indistinct triommatidia, ocular space 0.12 mm, interocular space 0.41 mm.

Antenna (Figure 152)

1.60 mm long, pale brown, 6 segmented, shorter than body; scape 0.08 mm long, 0.08 mm broad, smooth, dark brown with 2 stiff hairs; pedicel 0.06 mm long, 0.06 mm broad, smooth, dark brown with 2 stiff hairs; flagellum 4 segmented, pale brown, faintly imbricated with long, short hairs with acute apices; 1st segment 0.41 mm long 0.05 mm broad, pale brown, without secondary rhinaria, longest hair on 1st segment shorter, 0.45 x basal diameter of 1st segment; 2nd segment 0.23 mm long, 0.04 mm broad, without secondary rhinaria; 3rd segment 0.22 mm long, 0.03 mm broad; 4th segment 0.74 mm long, 0.02 mm broad, 4th segment base 0.25 mm long, 0.03 mm broad, dark brown; processus terminalis (p.t.) 0.49 mm long 0.02 mm broad, imbricated with serrated margin, dark brown; p.t. longer than base, 1.96 x base of last antennal segment; antenna shorter than body, 0.74 x body.

Flagellar Formula

1 L/W = 8.2, 2 L/W – 5.72, 4 L/W – 24.2, A = 12.70.

Rostrum (Figure 149, 153)

0.70 mm long, 0.04 mm broad, dark brown, extend beyond hindcoxa of metathoracic leg, upto middle of abdomen, 5 segmented; 4th and 5th segment fused to form ultimate rostral segment (u.r.s.), u.r.s. stout, 0.20 mm long, 0.03 mm broad, blackish, pointed at apex, 6 short hairs with acute apices; u.r.s. longer than h.t.2, 1.66 x h.t.2.

THORAX

0.59 mm long, 0.62 mm broad, pale brown, sparsely hairy, stiff hairs with acute apex, some with acuminate apices; broader than long; prothorax 0.17 mm long, 0.41 mm broad; mesothorax 0.20 mm long, 0.48 mm broad with distinct lateral tubercle; metathorax 0.23 mm long, 0.62 mm broad with distinct lateral tubercle.

Hind Leg (Figure 154)

1.29 mm long, pale brown, short, long hairs with acute, acuminate apices; coxa stout, 0.09 mm long, 0.08 mm broad, dark brown, smooth, sparsely hairy, stiff hairs with acute apices, broad at base, narrow at apex; trochanter 0.04 mm long, 0.06 mm broad, dark brown, smooth, sparsely hairy, stiff hairs with acute apices; femur 0.44 mm long, 0.08 mm broad, pale brown at base, dark brown at apex, ventrally with spinulose striae, dorsal hairs long; tibia long, narrow, 0.61 mm long, 0.04 mm broad, pale brown, dense, short, long hairs with acute, acuminate apices; tarsus 0.12 mm long, 0.02 mm broad, pale brown, 1st tarsal segment short, 0.02 mm long, pale brown, conical, with short 7 hairs; 2nd tarsal segment (h.t.2) 0.12 mm long, 0.02 mm broad, slightly curved, dark brown, faint imbrications with serrated margin with curved two claws; h.t.2 shorter than u.r.s., 0.60 x u.r.s.

ABDOMEN

1.20 mm long, 0.92 mm broad, dark brown, smooth, few oval pigmented spots, entirely spinulose, spinules only marginally, broad medially, oval with lateral tubercles; long, short hairs with acute, acuminate apices; 8 segmented; 1st segment 0.19 mm long, 0.48 mm broad; 2nd segment 0.20 mm long, 0.60 mm broad; 3rd segment 0.20 mm long, 0.82 mm broad; 4th segment 0.22 mm long, 0.92 mm broad; 5th segment 0.18 mm long, 0.80 mm broad; 6th segment 0.14 mm long, 0.60 mm broad; 7th segment 0.10 mm long, 0.10 mm broad; 8th segment 0.06 mm long, 0.11 mm broad; subanal plate round with 12 stiff hairs.

Siphunculus (Figure 155)

0.60 mm long, 0.10 mm broad, elongated, curved outward, dark brown, constricted at base, apex, broad at middle; inner margin of siphunculus straight but outer become convex, densely imbricated, nonserrated, hairs mostly uniform, long with fine apices; siphunculus longer than cauda, 0.5 x cauda, shorter than body, 0.30 x body.

Cauda (Figure 155)

0.12 mm long, 0.24 mm broad, short, broadly oval, pale brown, 8 long fine hairs with acute apices.

HOST PLANTS

Guava: *Pisidium guajava.*

DISCUSSION

Eutrichosiphum paratrichosiphum pisdiumi sp.nov. runs close to *Eutrichosiphum paratrichosiphum alnicola* Basu by having:

1. Antenna never less than 0.50 x body.
2. Processus terminalis about 1.40–2.10 x base of 6th segment.

However, differs from above species by following characters:

1. Lateral frontal tubercles on head absent, frons with weakly developed median frontal prominence.
2. Longest hair on 3rd segment 0.45 x basal diameter of 3rd segment of antenna.
3. Ultimate rostral segment dark with 6 short hairs with acute apices.
4. Subanal plate round with 12 stiff hairs.
5. Length of siphunculus six times to the width.
6. Flagellar formula: 1 L/W = 8.2, 2 L/W = 5.72, 4 L/W = 24.2, A = 12.70.

CONTROL MEASURES

Natural Control

Ladybird beetles: *Menochilus* sp., lacewing; hover fly; feed on aphids.

Chemical Control

Spray: pesticides as suggested for earlier species.

EUTRICHOSIPHUM NEOPARATRICHOSIPHUM RADHANAGARI SP. NOV.

FEMALE: APTERA (Figures 156–161)

1.92 mm long, 1.12 mm broad, dark brown; head 0.31 mm long, 0.48 mm broad; antenna 6 segmented, 1.29 mm long, shorter than body; thorax 0.36 mm long, 0.64 mm broad; abdomen 1.30 mm long, 1.12 mm broad; siphunculus 0.70 mm long, 0.10 mm broad, dark brown; cauda 0.14 mm long, 0.22 mm broad, pale brown.

HEAD (Figure 156)

0.31 mm long, 0.48 mm broad, dark brown, smooth, scabrous,

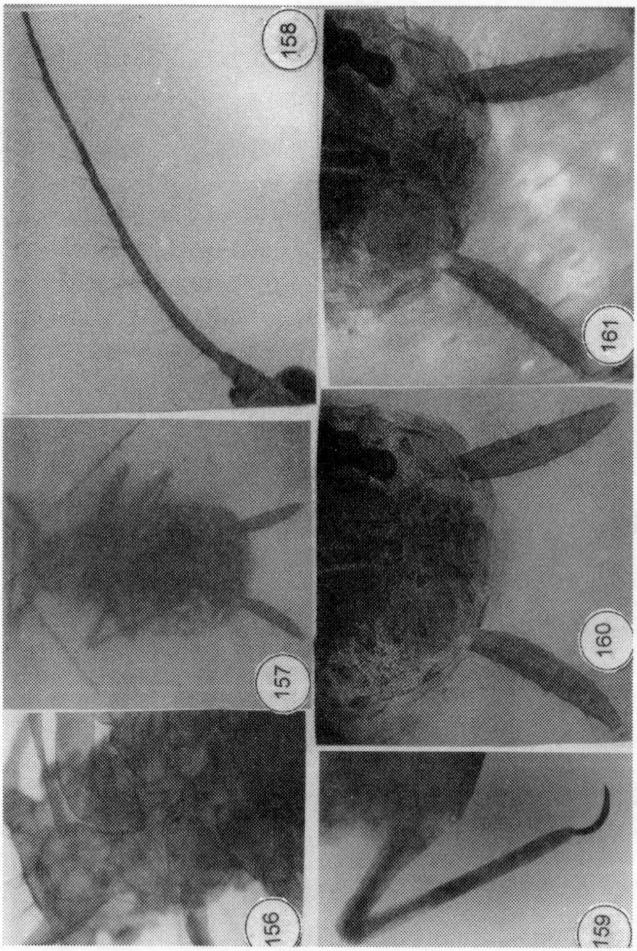

Figures 156-161: ***Eutrichosiphum neoparatrichosiphum radhanagari*** **sp.nov. (Female Aptera)**
Figure 156: Head and Rostrum; Figure 157: Whole Mount; Figure 158: Antenna; Figure 159: Hind Leg; Figure 160: Siphunculi; Figure 161: Cauda

long, stiff, fine hairs with acute and few with furcated apices; frons with weakly developed median frontal tubercle, lateral frontal tubercles absent; eyes lateral with indistinct triommatidia, ocular space 0.14 mm, interocular space 0.42 mm.

Antenna (Figure 158)

1.29 mm long, pale brown, 6 segmented, shorter than body; scape 0.10 mm long, 0.08 mm broad, dark brown, smooth, 2 long stiff, fine hairs with acute apices; pedicel 0.08 mm long, 0.08 mm broad, dark brown, smooth, 2 long stiff, fine hairs with acute apices; flagelium 4 segmented, pale brown, gradually more imbricated; 1st segment 0.25 mm long, 0.04 mm broad, without secondary rhinaria, 2nd segment 0.22 mm long, 0.04 mm broad; 3rd segment 0.21 mm long, 0.04 mm broad; 4th segment 0.35 mm long, 0.02 mm broad, slight dark brown, imbricated, 4th segment base 0.11 mm long, 0.03 mm broad; processus terminalis (p. t.) 0.26 mm long, 0.02 mm broad, dark brown, densely imbricated with serrated margin, processus terminalis longer than base, 2.0 x base of last antennal segment; antenna shorter than body, 0.67 x body.

Flagellar Formula

1 L/W= 7.5, 2 L/W = 5.5, 4 L/W = 11.35, A = 8.16.

Rostrum (Figure 156)

0.75 mm long, 0.04 mm broad, dark brown, extends beyond hindcoxa of metathoracic leg, upto anterior segment of abdomen, 5 segmented; 4th and 5th segment fused to form ultimate rostral segment (u.r.s.), u.r.s. 0.21 mm long, 0.03 mm broad, dark brown slight blackish with obtuse apex, u.r.s longer than h.t.2, 1.78 x h.t.2.

THORAX

0.36 mm long, 0.64 mm broad, dark brown, slight smooth, scabrous, broader than long, dorsum posses minute, stiff hairs with acute apices, few with furcated apices; prothorax 0.11 mm long, 0.48 mm broad; mesothorax 0.12 mm long, 0.52 mm broad; metathorax 0.10 mm long, 0.64 mm broad.

Hind Leg (Figure 159)

1.46 mm long, pale brown, short, long hairs with acute to acuminate apices; coxa stout, 0.10 mm long, 0.10 mm broad, dark brown, broad at base, tapered at apex, 3 stiff hairs with acute apices; trochanter short, straight 0.06 mm long, 0.06 mm broad, dark brown,

3 stiff hairs with acute apices; femur long, 0.40 mm long, 0.08 mm broad, dorsally distinct scattered spines, ventrally smooth; tibia long, narrow, 0.83 mm long, 0.04 mm broad, smooth with apical tuft of spines with indistinct imbrications except near apex of tibia; tarsus 0.07 mm long, 0.02 mm broad, pale brown; 1st tarsal segment short, 0.02 mm long, conical, pale brown, with 7 short hairs; 2nd tarsal segment(h.t.2) 0.07 mm long, 0.02 mm broad, slightly curved, dark brown, densely imbricated with serrated margin with curved two claws, hind tarsus (h.t.2) shorter than u.r.s, 0.32 x u.r.s.

ABDOMEN

1.30 mm long, 1.22 mm broad, oval, dark brown, with dark yellow pigmented round area throughout surface, dorsum with short and long hairs with acute apices, few with furcated apices, ventral spinulose at margin; 8 segmented; lateral abdominal tubercles indistinct on 1-7 segments; 1st segment 0.10 mm long, 0.82 mm broad; 2nd segment 0.20 mm long, 0.80 mm broad; 3rd segment 0.24 mm long, 0.96 mm broad; 4th segment 0.26 mm long, 1.12 mm broad; 5th segment 0.20 mm long, 0.92 mm broad; 6th segment 0.20 mm long, 0.72 mm broad; 7th segment 0.08 mm long, 0.70 mm broad; 8th segment 0.05 mm long, 0.30 mm broad; subanal plate round with 5 long, 5 short hairs with acute apices.

Siphunculus (Figure 160)

0.70 mm long, 0.10 mm broad, dark brown, elongated, outwardly directed, broad at middle, tapered at both base and apex, hairs long, stiff, fine on siphunculi with acute and furcated apices; siphunculus longer than cauda, 0.65 x cauda and shorter than body 0.36 x body.

Cauda (Figure 161)

Cauda 0.14 mm long, 0.22 mm broad, pale brown, transversely semioval with 6 hairs.

HOST PLANTS

Guava: *Pisidium guajava*.

DISCUSSION

This species runs close to *Eutrichosiphum Neoparatrichosiphum raychaudhuri*, Ghosh by having:

1. Processus terminalis about 1.80–2.00 x base of 6th segment.
2. Ultimate rostral segment about 1.60–1.95 x h.t. 2.

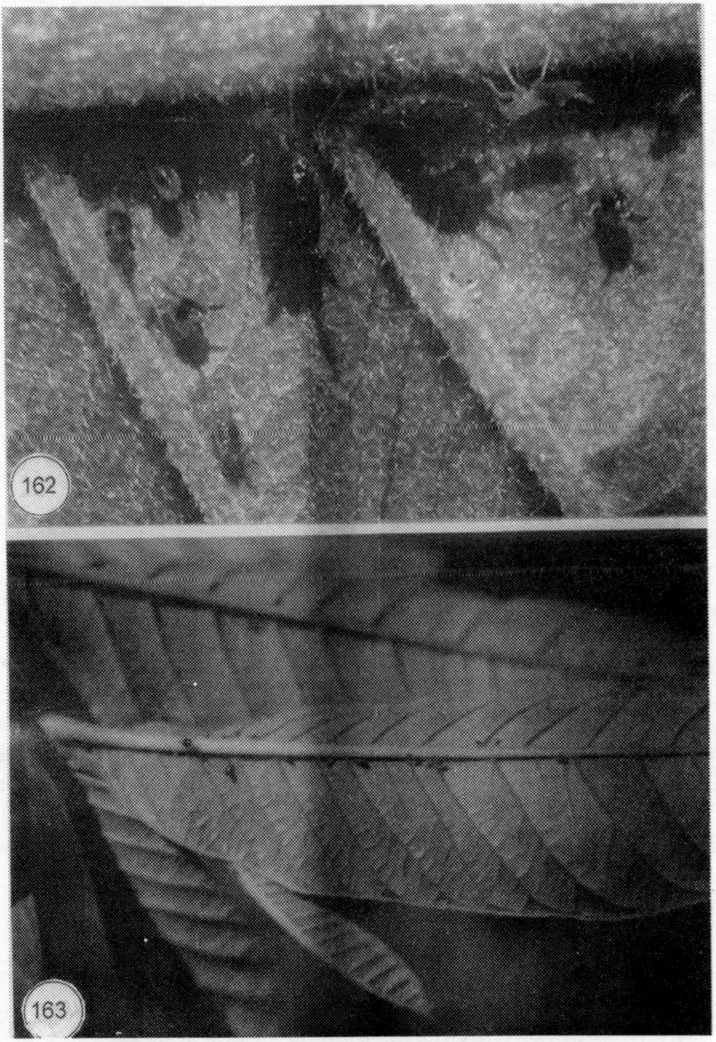

Figure 162: ***Eutrichosiphum paratrichosiphum pisidiumi*** **sp.nov.**
Figure 163: ***Eutrichosiphum neoparatrichosiphum radhanagari*** **sp.nov.**

However, differs from above species by following characters:

1. Processus terminalis longer than 3rd segment of antenna.
2. Ultimate rostral segment dark with obtuse apices.
3. 1st tarsal segment with 7 short hairs.
4. Dorsal abdominal hairs with acute and some with furcated apices.
5. Subanal plate with 5 long, 5 short hairs with acute apices.
6. Lateral frontal tubercles on head absent, frons with weakly developed median frontal prominence.
7. Flagellar formula: 1 L/W= 7.5, 2 L/W = 5.5, 4 L/W = 11.35, A = 8.16.

CONTROL MEASURES

Natural Control

Ladybird beetle, *Vernia* sp., *Brumoides* sp. and lace wings feed on the aphids.

Chemical Control

Spray: 0.03 per cent malathion (or) 0.03 per cent Azadirachtin (or) 0.02 per cent phosphamidon.

GENUS *GREENIDEA* SCHOUTENDEN

This genus was erected by Schoutenden in 1905, based on the type species *Siphonophora artocarpi* Westwood. The genus is widely distributed in India, Indonesia, Japan, Sri Lanka, Korea, Pakistan, Nepal, Malaysia, Taiwan, Thailand, Bangladesh, Australia, U.S.S.R. etc,

The genus shows following characteristics:

1. Body broadly oval or pear shaped.
2. Head smooth, without lateral frontal tubercles, frons straight or slightly convex, dorsal cephalic hairs usually long with furcated, acuminate or fine apices.
3. Antenna 6 segmented usually longer than body, flagellum mostly imbricated, sometimes reticulated (subgenus–*Neogreenidea*), p.t. longer than base of 6th segment, hairs on flagellum short or long with acute, acuminate or furcated apices; apterae without secondary rhinaria, alatae with

round to transversely oval secondary rhinaria on 3rd segment, primary rhinaria ciliated.

4. Rostrum long with distinctly divided segments 4 and 5; u.r.s. about 1.25 x 3.0 x h.t.2 and 4th segment with 8-16 secondary hairs.
5. Leg, femur and tibia smooth or with spinules, imbrications but in subgenus Greenidea hindtibia is with stradulatory ridges in the form of transverse cut. 1st tarsal segment with 7 ventral hairs.
6. Wings with normal venations.
7. Dorsum of abdomen pale brown to dark brown, smooth, wrinkled or corrugated and sometimes with some muscle plate like structure on pleural region, spinulose on dorsum at least anteriorly and laterally; dorsal hairs stiff, short and long, in alatae hairs mostly fine but sometimes acuminate apices.
8. Siphunculus elongated, about 0.20–1.01 x body in apterae, in alatae siphunculus always longer than apterae, curved outwards, densely covered with long hairs with acute, acuminate or furcated apices; in apterae reticulated pattern of transversely hexagonal cells either on both surfaces or only one surface restricted near the base, in alatae siphunculus always reticulated over entire length.
9. Cauda rounded or transversely oval with median stylus like process, sometimes small point or may be twice as long as wide at base, with few hairs.
10. Subanal plate broadly oval or semicircular.
11. Venter of abdomen rarely and densely spinulose, more usually minute spinulose striae present laterally on anterior sternites.

In India 28 species have been described from this genus.

Siphonophora Kach. Westwood, J. o. *Trans. Ent. Soc.*, London pt.4 (1890): 649.

Greenidea Schoutenden, H., Spolia Zeylan, 2 (1905); 181.

Trichosiphum Pergande, Th., Ent. News. Philad. 17 (1906): 207.

Greenidea schoutenden, H., Takahashi, R. Aph. of formosa, pt. 6 (1931): 30.

The genus *Greenidea* is subdivided into four subgenera namely:

1. *Trichosiphum*
2. *Neogreenidea*
3. *Paragreenidea*
4. *Greenidea.*

The subgenus *Trichosiphum* is characterised by following features.

1. Siphunculus in apterae distinctly reticulated on one surface at very base.
2. Siphunculus sometimes indistinctly reticulated so far beyond.

The subgenus *Neogreenidea* is recognized by following characteristics:

1. Siphunculus in apterae distinctly reticulated on both surfaces throughout length.
2. Flagellum distinctly reticulated and imbricated.

The subgenus *Paragreenidea* is visualised by following features:

1. Flagellum not reticulated but with normal imbrications.
2. Hind tibia with stradulatory ridges in the form of transverse cut.

The subgenus *Greenidea* is featured by:

1. Siphunculus in apterae distinctly reticulated on both surfaces throughout the length.
2. Hind tibia without any transverse cut.

From India 9 species of *Greenidea* have been described.

GREENIDEA GREENIDEA FHANASAE SP.NOV.

FEMALE: APTERA (Figures 164–169)

1.95 mm long, 0.95 mm broad, pale brown; head smooth, 0.30 mm long, 0.48 mm broad; antenna 6 segmented, 2.20 mm long, longer than body; thorax 0.62 mm long, 0.68 mm broad; abdomen 1.30 mm long, 0.95 mm broad; siphunculus distinctly reticulated, 1.10 mm

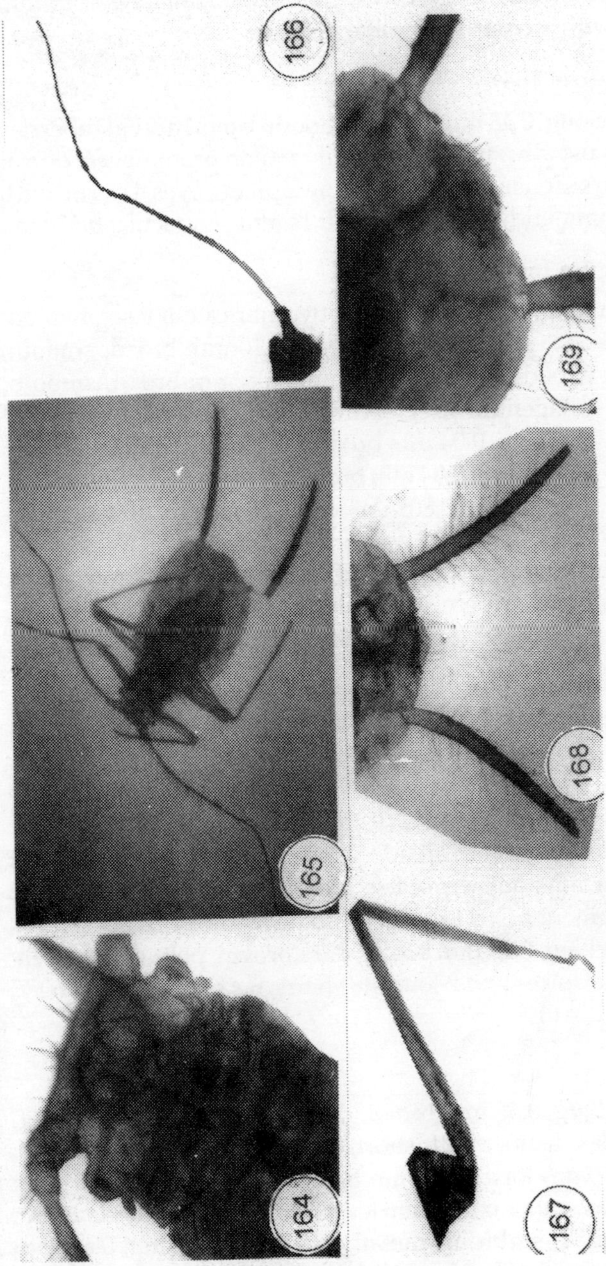

Figures 164-169: *Greenidea fhanasae* sp.nov. (Female Aptera)
Figure 164: Head and Rostrum; Figure 165: Whole Mount; Figure 166: Antenna;
Figure 167: Hind Leg; Figure 168: Siphunculi; Figure 169: Cauda

long, 0.09 mm broad, dark brown; cauda 0.07 mm long, 0.15 mm broad, oval with median stylus like process.

HEAD (Figure 164)

0.30 mm long, 0.48 mm broad, smooth, lateral frontal tubercles absent; frons usually straight without median prominence, dorsal cephalic hairs dense, long with fine apices; eyes lateral with indistinct triommatidia, ocular space 0.14 mm, interocular 0.41 mm.

Antenna (Figure 166)

2.20 mm long, pale brown, mostly imbricated, 6 segmented, longer than body; scape 0.10 mm long, 0.10 mm broad, smooth, sparsely hairy; pedicel 0.06 mm long, 0.08 mm broad, smooth, sparsely hairy; flagellum 4 segmented, imbricated, pale brown; 1st segment 0.64 mm long, 0.04 mm broad, secondary rhinaria absent; 2nd segment 0.30 mm long 0.04 mm broad; 3rd segment 0.38 mm long, 0.03 mm broad; 4th segment 0.80 mm long, 0.02 mm broad, 4th segment base long, 0.26 mm long, 0.02 mm broad, pale brown, imbricated, primary rhinarium ciliated at apex of base; processus terminalis (p. t.) 0.54 mm long, pale brown, imbricated with serrated margin, long, short hairs with acute apices; p.t. longer than base, 2.07 x base; longer than body, 1.28 x body.

Flagellar Formula

1 L/W= 16, 2 L/W= 7.5, 4 L/W= 26, A = 16.5.

Rostrum (Figure 164)

0.72 mm long, 0.04 mm broad, dark brown, long, extend beyond hindcoxa of metathoracic leg, upto middle of abdomen, 5 segmented; 4th and 5th segment fused to form ultimate rostral segment (u.r.s.), u.r.s. 0.23 mm long, 0.04 mm broad, dark brown, pointed, sparsely hairy with fine apices and 2 long stiff hairs; u.r.s. longer than h.t.2, 2.3 x h.t.2.

THORAX

0.62 mm long, 0.68 mm broad, dark brown, as broad as long, lateral tubercles distinct with short, long hairs with acute apices; prothorax 0.18 mm long, 0.48 mm broad, pronotum of prothorax fused to head; mesothorax 0.24 mm long, 0.65 mm broad, mesonotum, metanotum fused, scabrous; metathorax 0.24 mm long, 0.68 mm broad.

Hind Leg (Figure 167)

1.60 mm long, pale brown; coxa stout, 0.11 mm long, 0.14 mm broad, dark brown, broad at base, narrow at apex with sparsely hairy; trochanter stout, 0.06 mm long 0.08 mm broad, dark brown, sparsely hairy; femur 0.70 mm long, 0.07 mm broad, pale brown, imbricated, vertical lines with moderate short and long hairs; tibia long, narrow, 0.69 mm long, 0.03 mm broad, short, long hairs, dense towards apex, hind tibia without any transverse cut; tarsus 0.12 mm long, 0.02 mm broad; 1st tarsal segment very short, 0.02 mm long, pale brown; 2nd tarsal segment (h.t.2) 0.10 mm long, 0.02 mm broad, faintly imbricated, slightly curved with curved two claws, hind tarsus (h.t.2) shorter than u.r.s., 0.44 x u.r.s.

ABDOMEN

1.30 mm long, 0.95 mm broad, pale brown, smooth, wrinkled, muscle patch like pleural region, spinulose on dorsum anteriorly, laterally, dorsal hairs stiff, with acute apices, venter rarely spinulose, minute striae laterally on anterior sternite; 8 segmented; lateral abdominal tubercles distinct; 1st segment 0.18 mm long, 0.80 mm broad; 2nd segment 0.18 mm long, 0.82 mm broad; 3rd segment 0.22 mm long, 0.88 mm broad; 4th segment 0.22 mm long, 0.95 mm broad; 5th segment 0.20 mm long, 0.92 mm broad; 6th segment 0.16 mm long, 0.86 mm broad; 7th segment 0.14 mm long, 0.80 mm broad; 8th segment 0.09 mm long, 0.65 mm broad; subanal plate broadly oval.

Siphunculus (Figure 168)

1.10 mm long, 0.09 mm broad, elongated, curved outward, dark brown, dense long hairs with acute apices, distinctly reticulated on dorsal, ventral surface throughout length; siphunculus narrow at base, broad medially and tapered at apex; siphunculus shorter than body, 0.56 x body.

Cauda (Figure 169)

Cauda short, 0.07 mm long, 0.15 mm broad, pale brown, oval with median stylus like process, 8 long stiff hairs, with acute apices.

HOST PLANTS

Jackfruit: *Artocarpus integrifolia.*

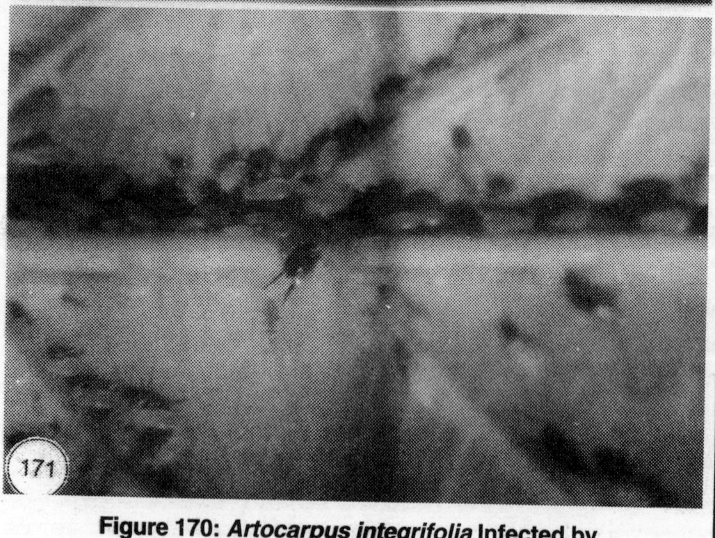

Figure 170: ***Artocarpus integrifolia*** **Infected by *Greenidea Greenidea fhanasae* sp.nov.**

Figure 171: ***Greenidea Greenidea fhanasae*** **sp.nov.**

DISCUSSION

This species runs close to *Greenidea Greenidea photiniphaga*, Raychaudhuri and Ghosh by having following characters:

1. Siphunculi longer than hind tibiae.
2. Venter area free from spinulose.

However, differs from having following characters:

1. Dorsum of abdomen smooth, locally sclerotized.
2. Dorsum of abdomen pale brown.
3. Cauda with more than 8 hairs, 10–12.
4. Body 1.95 mm long, 0.90 mm broad.
5. Antenna longer, 2.20 mm long.
6. u.r.s. 0.23 mm long upto about 2.30 mm x as long as h.t.2.
7. Siphunculus shorter, 1.10 mm.
8. Flagellar formula: 1L/W = 16, 2 L/W = 7.5, 4 L/W = 26, A = 16.5.

CONTROL MEASURES

Natural Control

Ladybird beetles: *Menochilus* sp., *Coccinella* sp., lacewing, hover fly, feed on aphids.

Chemical Control

As above species.

GREENIDEA GREENIDEA BHIMASHANKARI SP.NOV.

FEMALE: APTERA (Figures 172–176)

1.90 mm long, 1.12 mm broad, dark brown; head 0.31 mm long, 0.48 mm broad; antenna 2.29 mm long, 6 segmented, longer than body; thorax 0.36 mm long, 0.64 mm broad; abdomen 1.20 mm long, 1.11 mm broad; siphunculus 0.70 mm long, 0.10 mm broad, dark brown; cauda short 0.12mm long, 0.20 mm broad, pale brown.

HEAD (Figure 172)

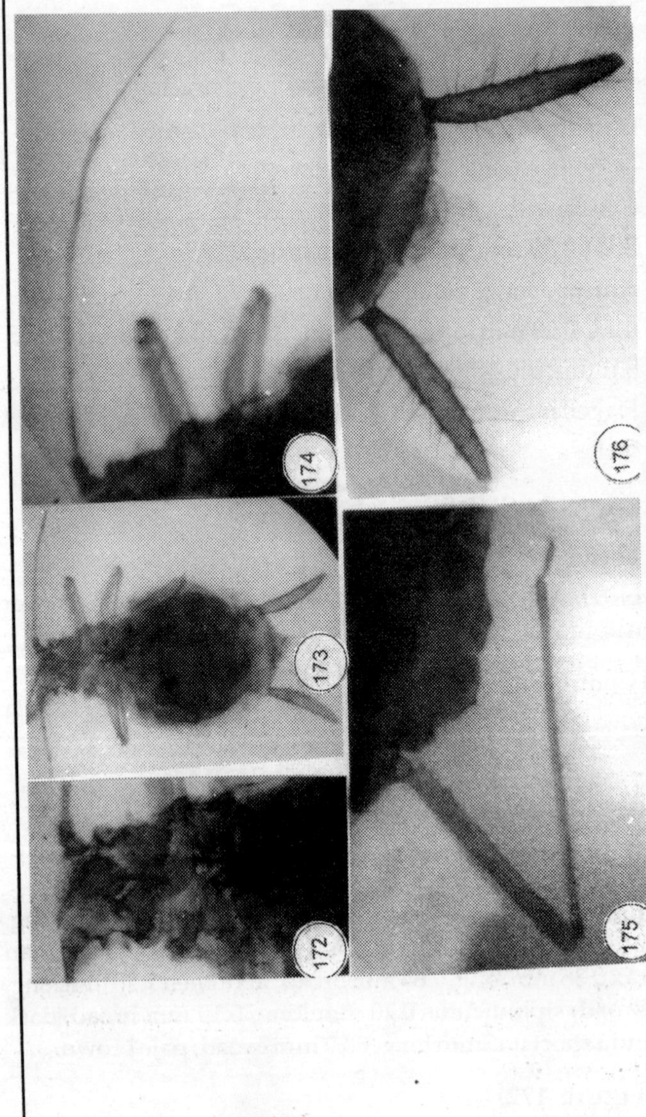

Figures 172-176: *Greenidea bhimashankari* sp.nov. (Female Aptera)
Figure 172: Head and Rostrum; Figure 173: Whole Mount; Figure 174: Antenna; Figure 175: Hind Leg; Figure 176: Siphunculi and Cauda

0.31 mm long, 0.48 mm broad, dark brown, head smooth; frons slightly convex, lateral frontal tubercles absent; dorsal cephalic hairs moderate, long with acuminate apices; eyes lateral with distinct triommatidia, ocular space 0.08 mm, interocular space 0.42 mm.

Antenna (Figure 174)

2.29mm long, slight pale brown, mostly imbricated; 6 segmented, longer than body; scape stout 0.10 mm long, 0.08 mm broad, dark brown, sparsely hairy with acuminate apices; pedicel short, 0.08 mm long, 0.06 mm broad, dark brown; sparsely hairy with acuminate apices; flagellum long, 4 segmented, pale brown; 1st segment 0.40 mm long, 0.04 mm broad, densely imbricated, without secondary rhinaria; 2nd segment 0.31 mm long, 0.04 mm broad; 3rd segment 0.26 mm long, 0.04 mm broad; 4th segment 0.35 mm long, pale brown, densely imbricated with serrated margin, sparsely hairy, 4th segment base 0.11 mm long, 0.03 broad, primary rhinarium ciliated at tip of base; processus terminalis (p.t.) 0.24 mm long, 0.02 mm broad, pale brown, imbricated; p.t. longer than base, 2.18 x base of last antennal segment; antenna longer than body, 1.20 x body.

Flagellar Formula

1 L/W = 10.00, 2L/W = 5.5, 4 L/W = 11.66, A = 9.05.

Rostrum (Figure 172)

0.75 mm long, 0.04 mm broad, dark brown, extends beyond hindcoxa of midthoracic leg, upto middle of abdomen, 5 segmented; 4th and 5th segment fused to form ultimate rostral segment (u.r.s.), u.r.s. 0.21 mm long, 0.04 mm broad, dark blackish, pointed apex, 12 short hairs with acute apices; u.r.s. longer than h.t. 2, 3.0 x h.t.2.

THORAX

0.36 mm long, 0.64 mm broad, dark brown, broader than long, lateral tubercles weakly developed, dorsum smooth with sparsely hairy; prothorax 0.14 mm long, 0.48 mm broad; mesothorax 0.12 mm long, 0.52 mm broad; metathorax 0.10 mm long, 0.64 mm broad; mesothorax and metathorax fused together.

Hind Leg (Figure 175)

1.46 mm long, pale brown, smooth, imbricated; coxa stout 0.10 mm long, 0.10 mm broad, broad at base, narrow apex, as long as broad, dark brown, sparsely hairy; trochanter short, straight, 0.06mm long, 0.06mm broad, dark brown, sparsely hairy; fumer long, smooth and densely imbricated 0.40 mm long, 0.06 mm broad; tibia 0.83 mm long, 0.04 mm broad, densely imbricated, densely hairy at apex than base with acuminate apices; hind tibia 0.40 mm tarsus 0.07mm long, 0.02 mm broad; 1st tarsal segment short, 0.02 mm long, pale brown with 6 ventral hairs; 2nd tarsal segment (h.t.2) 0.07 mm long, slightly curved with curved two claws; 2nd tarsal segment (h.t.2.) shorter than u.r.s, 0.33 x u.r.s.

ABDOMEN

1.20 mm long, 1.11 mm broad, dark brown, smooth, slightly wrinkled, densely spinulose ventrally, dorsal hairs long, short with acute and some with bifurcated apices; 8 segmented; lateral abdominal tubercles distinct on 1-7 segments; 1st segment 0.10 mm long, 0.82 mm broad; 2nd segment 0.20 mm long, 0.90 mm broad; 3rd segment 0.24 mm long, 0.96 mm broad; 4th segment 0.26 mm long, 1.11 mm broad; 5th segment 0.20 mm long, 0.02 mm broad; 6th segment 0.20 mm long, 0.72 mm broad; 7th segment 0.08 mm long, 0.60 mm broad and 8th segment 0.05 mm long, 0.30 mm broad, subanal plate semicircular.

Siphunculus (Figure 176)

0.70 mm long, 0.10 mm broad, dark brown, elongated, densely hairy with acute and bifurcated apices, distinctly imbricated throughout long; siphunculi longer than cauda, 5.10 x cauda, shorter than body, 0.63 x body.

Cauda (Figure 177)

0.12 mm long, 0.20 mm broad, pale brown, oval; with 4 primary, 4 secondary hairs, stiff with acute apices.

HOST PLANTS

Quercus sp; Guava: *Pisdium guajava.*

DISCUSSION

This species runs close to *Greenidea Greenidea decaspermi* Takahashi by having:

1. Siphunculi longer than hind tibia.
2. Ultimate rostral segment larger than h.t. 2.

However, differs from above species by following characters:

1. Dorsal abdominal hairs with acute and bifurcated apices.
2. Siphunculus distinctly imbricated throughout length.
3. Lateral tubercles on head absent, frons with slight convex median frontal tubercle.
4. Ultimate rostral segment, pointed apex with 12 short hairs.
5. Cauda with 4 primary, 4 secondary hairs with acute apices.
6. Flagellar formula: 1 L/W = 10.00, 2 L/W = 5.5, 4 L/W = 11.66, A = 9.05.

CONTROL MEASURES

Natural Control

Ladybird beetles: *Illeis* sp., *Micraspis* sp., feed on aphids.

Chemical Control

Spray: 0.1 per cent Carbaryl (or) 0.03 per cent DDVP (or) 0.03 per cent phosphamidon (or) 0.03 per cent malathion.

GENUS *CERATOVACUNA* ZEHNTNER

This genus was erected by Zehntner in 1897 with the help of type species *Ceratovacuna lanigera* Zehntner which is cosmopolitan and widely distributed throughout India. It has been reported from Assam, Sikkim, Tripura, Uttar Pradesh and West Bengal. Outside India, the genus *Ceratovacuna* has been reported from Indonesia, Japan, Philippines and Sri Lanka. The genus shows following characteristics:

1. Body oval or pear shaped.
2. In apterae head and prothorax fused, with pair of short to moderately developed bluntish horns with few hairs, in alatae much reduced; cephalothorax in apterae with group of wax gland cells become pale to smoky; dorsal cephalic hairs usually long with acute or swollen apices or both.
3. Antenna 4 or 5 segmented in apterae while 5 segmented in alatae; short, about 0.10–0.40 x body; flagellum imbricated may dense or sparse, flagellar hairs may short or fairly long, secondary rhinaria only in alatae, primary rhinaria round and ciliated, p.t. short, but about 0.50 x base of last antennal segment.

4. Rostrum usually short, extend upto midcoxa of mesothoracic leg, u.r.s. blunt, short, about 0.40–0.80 x h.t. 2, without secondary hairs.
5. Legs smooth or minute spinulose, F.T.C. 3,3,2, 3,2,2 or 4,3,2, 2nd tarsal segment with a pair of dorsoapical hairs with funnel shaped apices.
6. Wings-forewing with M once branched; pterostigma short, blunt distally, hindwings with two oblique veins.
7. In apterae abdomen dorsum pale or may sclerotic around wax gland cells, may arranged marginally, pleurally, dorsum hairs, minute, sparse, with acute apices or have swollen apices while alatae variably developed, spinopleural sclerotic dorsal bands on post siphuncular segment.
8. Siphunculus poriform with sclerotic rim.
9. Cauda with basal constriction, knobbed.
10. Subanal plate bilobed.
11. Subgenital plate normal.

From India, five species have been reported under this genus.

CERATOVACUNA BAMBUSI SP.NOV.

FEMALE: APTERA (Figures 177–181)

1.85 mm long, 0.98 mm broad, greenish with white patches; head 0.23 mm long, 0.36 mm broad; antenna 5 segmented, 0.48mm long, shorter than body; thorax 0.42 mm long, 0.68 mm broad; abdomen 1.05 mm long, 0.98 mm broad; siphunculus with siphuncular pore; cauda weakly developed, 0.02 mm long, 0.08 mm broad.

HEAD (Figure 177)

0.23 mm long, 0.36 mm broad, dark greenish, head, prothorax, fused to form short bluntish horn with few minute hairs; wax gland cells in groups, pale green; dorsal cephalic hairs long with acute apices, 2.8 x basal diameter of 3rd segment; frons become flat, without median prominence, lateral frontal tubercles very weakly developed; eyes lateral with indistinct triommatidia, ocular space 0.12 mm, interocular space 0.32 mm.

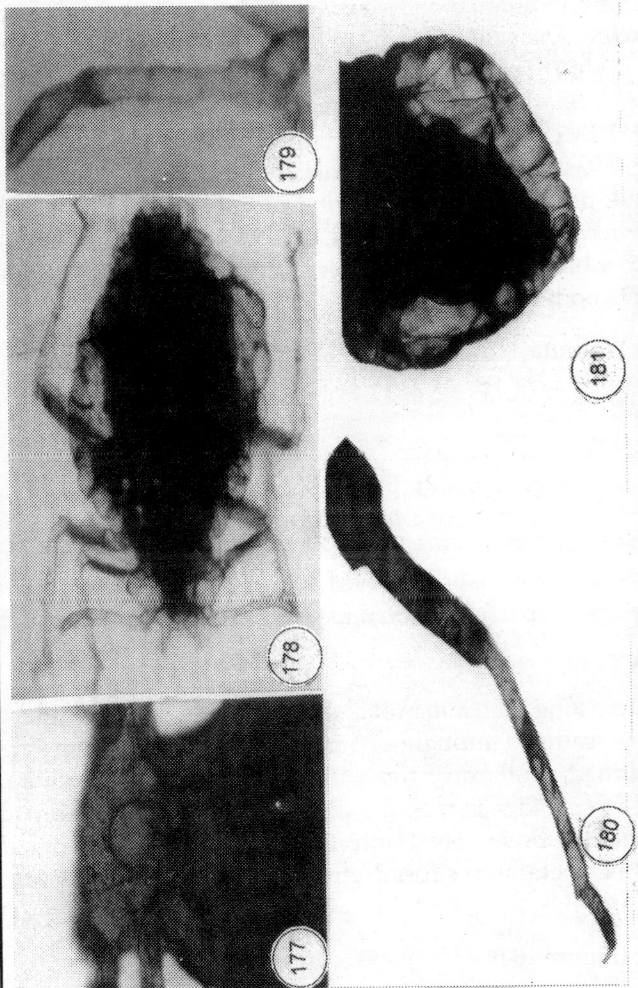

Figures 177-181: *Ceratovacuna bambusi* sp.nov. (Female Aptera)
Figure 177: Head and Rostrum; Figure 178: Whole Mount; Figure 179: Antenna;
Figure 180: Hind Leg; Figure 181: Siphunculi and Cauda

Antenna (Figure 179)

0.48 mm long, pale greenish, imbricated, 5 segmented, shorter than body; scape stout, 0.06 mm long, 0.06 mm broad, dark greenish, faintly imbricated, sparsely hairy with acute apices; pedicel short, 0.02 mm long, 0.02 mm broad, dark greenish, faintly imbricated, as long as wide, sparsely hairy; flagellum 3 segmented, imbricated, pale greenish; 1st segment 0.18 mm long, 0.04 mm broad, without secondary rhinaria, longest hair on 3rd segment, 1.20 x its basal diameter; 2nd segment 0.11 mm long 0.03 mm broad; 3rd segment 0.10 mm long, 0.02 mm broad, 3rd segment base short, 0.08 mm long, 0.02 mm broad, primary rhinarium round protuberant at tip of base; processus terminalis (p.t.) very short, 0.02 mm long; p.t. shorter than base, 0.25 x base of last antennal segment; antenna shorter than body, 0.29 x body.

Flagellar Formula

1 L/W= 4.5, 2 L/W= 5.4, 3 L/W = 5.0, A = 4.38.

Rostrum (Figure 177)

0.20 mm long, 0.07 mm broad, short, dark greenish, extend between forecoxa, midcoxa of prothoracic, mesothoracic leg, 5 segmented; 4th and 5th segment fused to form ultimate rostral segment (u.r.s), u.r.s dark greenish, very short, 0.05 mm long, 0.04 mm broad with blunt apex, sparsely hairy, u.r.s shorter than h.t.2, 0.62 x h.t.2.

THORAX

0.42 mm long, 0.68 mm broad, dark greenish, oval, wax gland cells, shiny, scattered throughout dorsum; prothorax 0.12 mm long, 0.34 mm broad, prothorax fuses with head to form cephalothorax with oval wax gland cells; mesothorax 0.19 mm long, 0.48 mm broad, wax gland cells absent; metathorax 0.12 mm long, 0.68 mm broad; mesothorax, metathorax fused with weakly developed lateral tubercles.

Hind Leg (Figure 180)

0.74 mm long, dark greenish, smooth, with faintly spinulose; coxa 0.06 mm long, 0.06 mm broad, dark greenish, broad at base, narrow at apex; trochanter 0.04 long, 0.04 mm broad, dark greenish, slightly spinulose; femur long, 0.18 mm long, 0.04 mm broad, pale at the base, dark at apex, hairs short, moderate with acute apices, longest hair on femora shorter, 0.50 x basal diameter of 3rd segment; tibia long, narrow, 0.37 mm long, 0.02 mm broad, densely spinulose with short, long hairs with acute apices; tarsus short, 0.08 mm long, 0.02 mm broad, 1st tarsal segment short, 0.01 mm long, pale greenish, F. T. C. with 3,3,2, fine hairs; 2nd tarsal segment (h.t.2) 0.08 mm long, slightly curved with curved two claws; h.t.2 longer than u.r.s., 1.06 x u.r.s.

ABDOMEN

1.05 mm long, 0.98 mm broad, greenish with white patches, sclerotic, spinal pleural cell groups absent, wax gland cells in groups arranged in anterior tergites, wax gland cells arranged irregularly on 8th tergite abdomen, sparsely hairy with acute apices; 8 segmented; lateral abdominal tubercles weakly developed on segments 1-7; 1st segment 0.14 mm long, 0.80 mm broad; 2nd segment 0. 12 mm long, 0.84 mm broad; 3rd segment 0.14 mm long, 0.89 mm broad; 4th segment 0.14 mm long, 0.80 mm broad; 6th segment 0.14 mm long, 0.65 mm broad; 7th segment 0.11 mm long, 0.45 mm broad; 8th segment 0.07mm long, 0.30 mm broad; subgenital plate normal, oval, sparsely hairy with acute apices; subanal plate bilobed, pale brown.

Siphunculus (Figure 181)

Siphunculus poriform, very short, 0.05 mm, it often with sclerotic rim.

Cauda (Figure 181)

0.02 mm long, 0.08 broad, very short, weakly developed, basal constriction, semi oval, with 5 short hairs.

HOST PLANTS

Bamboo: *Bamboo* sp.

DISCUSSION

This species runs close to *Ceratovacuna spinulosa* Ghosh and Raychaudhuri by:

1. Dorsum of abdomen with spinal wax gland cell groups on anterior segment.
2. Wax gland cells on 8th tergite arranged irregularly.
3. Flagellum with dense spinulose imbrications.

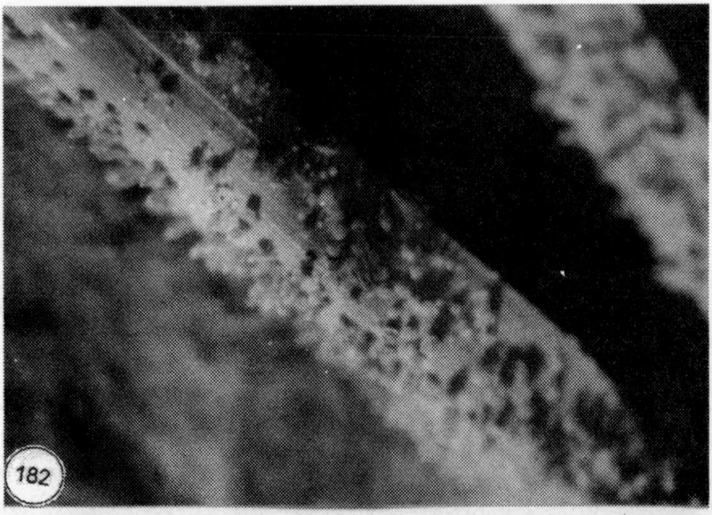

Figure 182: Bamboo Infected by *Ceratovacuna bambusi* sp.nov.
Figure 183: Colony of *Ceratovacuna bambusi* sp.nov.

However, differs from above species by following characters:

1. Dorsal cephalic hairs longer, 2.8 x basal diameter 3rd segment.
2. Lateral frontal tubercles on head weakly developed, frons without median frontal prominence.
3. Subgenital plate normal, oval, sparsely hairy.
4. Cauda very short, semioval with 5 short hairs.
5. Longest hair 3rd segment longer, 1.20 x basal diameter of 3rd segment.
6. Flagellar formula: 1 L/W= 4.5, 2 L/W= 5.4, 3 L/W = 5.0, A = 4.38.

CONTROL MEASURES

Natural Control

Ladybird beetle, *Vernia* sp., Tiger beetle *Eudema angulatum*, feed on aphids.

Chemical Control

Spray: 0.1 per cent carbaryl (or) 0.03 per cent phosphamidon (or) 0.05 per cent malathion (or) 0.03 per cent DDVP.

CERATOVACUNA LANIGERA ZEHNTNER

FEMALE: ALATA (Figures 184–195)

2.25 mm long, 1.18 mm broad, dark brownish, oval; head 0.21 mm long, 0.49 mm broad; rostrum very short; antenna 5 segmented, 0.95 mm long, shorter than body; thorax 0.76 mm long, 0.90 mm broad; abdomen 1.45 mm long, 1.18 mm broad, dark brownish; siphunculus poriform with thick sclerotic rim.

HEAD (Figure 188)

0.21 mm long, 0.49 mm broad, broader than long, dark brownish, head and prothorax fused, small paired horn placed quite apart just near lateral frontal tubercles; dorsal cephalic hairs long with acute apices; well defined wax gland cells pale to smoky; eyes lateral with distinct triommatidia, ocular space 0.14 mm, interocular space 0.40 mm.

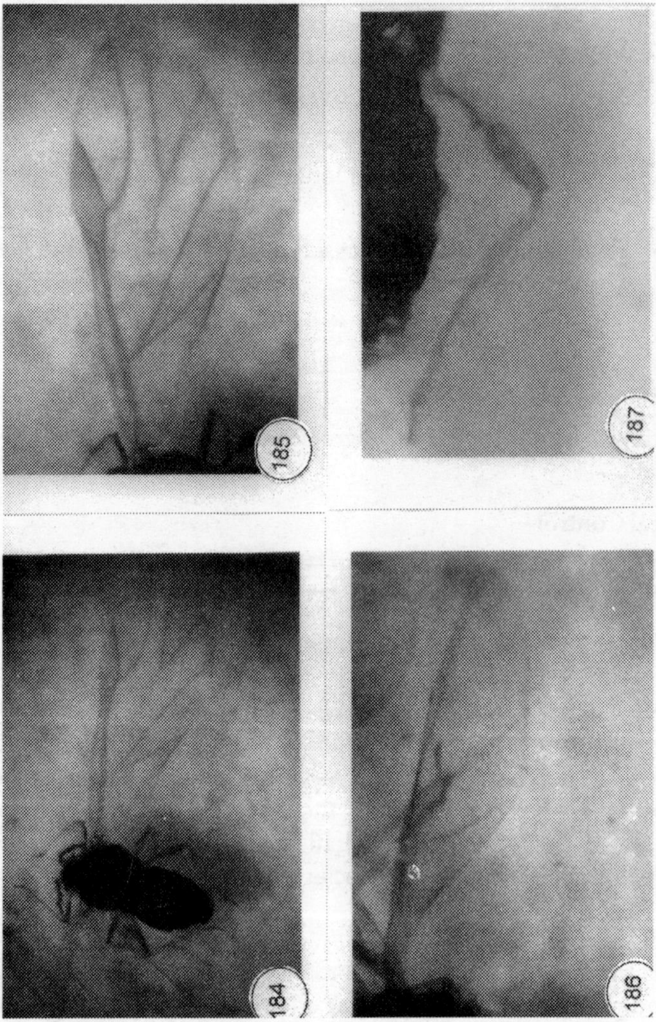

Figures 184-187: *Ceratovacuna lanigera* (Female Aptera)
Figure 184: Female Alata; Figure 185: Fore Wing; Figure 186: Hind Wing; Figure 187: Hind Leg

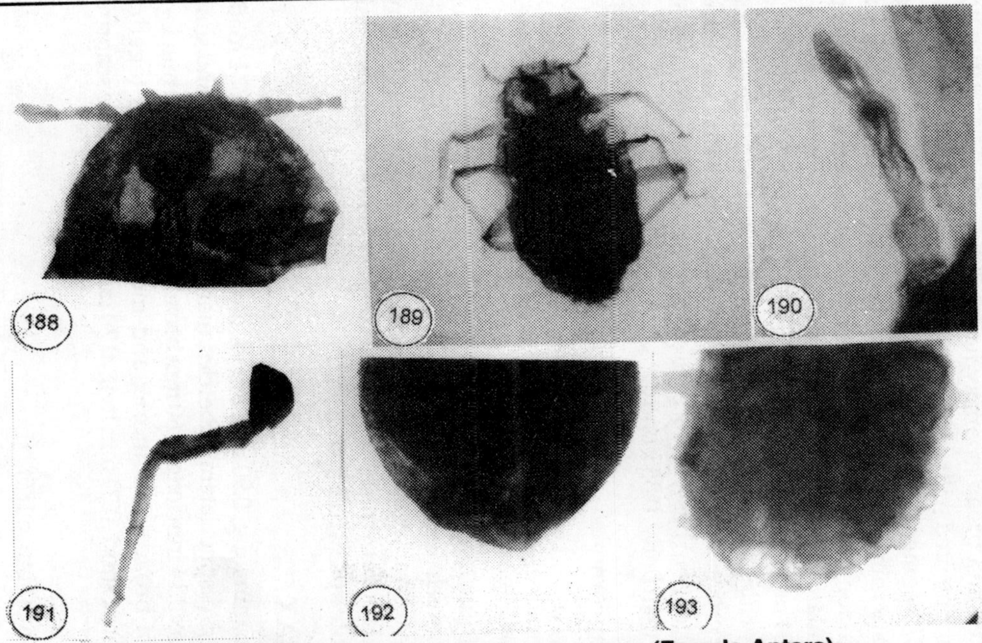

Figures 188-193: *Ceratovacuna lanigera* sp.nov (Female Aptera)
Figure 188: Head and Rostrum; Figure 189: Whole Mount; Figure 190: Antenna; Figure 191: Hind Leg; Figure 192: Siphuncular Pore; Figure 193: Cauda

Antenna (Figures 190)

0.80 mm long, 0.04 mm broad, dark brown, 5 segmented, shorter than body; scape stout, 0.10 mm long, 0.08 mm broad, smooth, sparsely hairy with acute apices; pedicel short, straight, 0.09 mm long, 0.07 mm broad, sparsely hairy with acute apices; flagellum long, 3 segmented with well serrated transverse ridges with sparsely hairy; 1st segment 0.36 mm long, 0.08 mm broad, smooth, many secondary rhinaria become subannular, closely placed; 2nd segment 0.20 mm long, 0.07 mm broad, many secondary rhinaria become subannular, sparsely hairy with acute apices; 3rd segment 20 mm long, 0.04 mm broad, 3rd segment, base 0.14 mm long, 0.02 mm broad, primary rhinarium round, slightly protuberant with fine cilia at tip of base; processus terminalis (p.t.) 0.06 mm long with 4 short hairs; p.t. shorter than base, 0.42 x base of last antennal segment; antenna shorter than body, 0.35 x body.

Flagellar Formula

1 L/W= 5.14, 2 L/W= 2.85, 3 L/W= 5.00, A= 4.33.

Rostrum (Figure 188)

0.15 mm long, 0.08 mm broad, very short, extends upto forecoxa of prothoracic leg, 5 segmented; 4th and 5th segment fused to form ultimate rostral segment (u.r.s.); u.r.s. short, 0.09 mm long, 0.08 mm broad, dark brownish, without secondary hairs, 2 primary hairs with acute apices; u.r.s. shorter than h.t. 2, 0.79 x h.t. 2.

THORAX

0.76 mm long, 0.90 mm broad, dark brown, scabrous, broader than long, distinct lateral tubercles, many wax pores oval and shiny; prothorax 0.14 mm long, 0.60 mm broad; mesothorax 0.40 mm long, 0.90 mm broad, mesothorax and metathorax fused together; metathorax short, 0.24 mm long, 0.70 mm broad, become scabrous, sclerotic.

Fore Wing (Figure 185)

3.05 mm long, 1.30 mm broad, membranous, longer than body; consists pterostigma vein, redial sector, media, cubitus and anal vein; pterostigma distinct, short and bluntish distally, media once branched, radial sector curved and touch to pterostigma; cubitus and anal vein straight, originated from same point become 'V' shaped; all veins distinct, dark brown.

Figure 194: Sugarcane Infected by *Ceratovacuna lanigera*
Figure 195: Colony of *Ceratovacuna lanigera*

Hind Wing (Figure 186)

Hind wing short, membranous 1.60 mm long, 0.61 mm broad; two oblique veins (cubitus and anal vein) simple straight, parallel to each other; veins distinct, dark brown.

Hing Leg (Figure 187, 191)

1.45 mm long, dark brown, smooth, sparsely hairy; coxa stout, 0.10 mm long, 0.08 mm broad, smooth, without hairs, broad at base, narrow to apex; trochanter short, triangular 0.06 mm long, 0.06 mm broad, smooth, without hairs; femur 0.48 mm long, 0.08 mm broad, straight, stout, broad at apex; tibia 0.66 mm long, 0.06 mm broad with primary, secondary hairs (spines), dense with acute apices; tarsus 0.19 mm long, 0.04 mm broad; 1st tarsal segment short, conical, pale brown, F.T.C. with 3,3,3; 2nd tarsal segment slightly curved, imbricated with curved two claws with pair of dorsoapical hairs with funnel shaped apices; hind tarsus (h.t. 2) longer than u.r.s., 1.26 x u.r.s.

ABDOMEN

1.45 mm long, 1.18 mm broad, dark brown, smooth, slight scabrous with numerous wax pores; lacking spinal and plural wax gland cell groups, well defined wax glands composed of cells lying close to each other in groups and placed marginally, dorsal hairs sparse with acute apices; 1st segment 0.22 mm long, 0.80 mm broad; 2nd segment 0.22 mm long, 0.90 mm broad; 3rd segment 0.20 mm long, 0.99 mm broad; 4th segment 0.20 mm long, 1.18 mm broad; 5th segment 0.18 mm long, 0.95 mm broad; 6th segment 0.18 mm long, 0.85 mm broad; 7th segment 0.16 mm long, 0.70 mm broad; 8th segment slight hexagonal, small and distinct, 8th tergite wax gland cells arranged irregularly, sparsely hairy with acute apices, subgenital plate normal, subanal plate bilobed.

Siphunculus (Figure 192)

Siphunculus become poriform, minute pores, siphuncular pore 0.05 mm long.

Cauda (Figure 193)

Cauda short with basal constriction and knobbed.

HOST PLANTS

Sugarcane *Saccharum officinarum*.

CONTROL MEASURES

Natural Control

Ladybird beetles: *Menochilus* sp., *Coccinella* sp., lacewing, hover fly, Syphrid, *Conobatra aphidivora*, etc. feed on aphids and suppress aphid populations.

Chemical Control

As suggested under Chapter 3.

Bibliography

Agarwala, B. K. 1980. Some aspects of aphid (Homoptera: Insecta) studies in Sikkim and Bhutan. Ph. D. thesis Zoology, Calcutta Uni., pp. 383.

Agarwala, B. K. and A. K. Ghosh, 1984. A check list of Aphidoidea of India. *Rec. Zool. Surv. India*, 50, 1–71.

Agarwala, B. K. and A. K. Ghosh, 1988. Prey records of aphidophagous Coccinellidae in India. A review and bibliography. *Trop. Pest. Manage*, 34, 1–14.

Agarwala, B. K. and J. H. Martin, 1993. A taxonomic study of the genus *Anomalosiphum* Takahashi 1934 (Homoptera: Aphididae: Greenideinae). *J. Linn. Soc.* (Zool.), 45–49.

Agarwala, B. K., Mondal, P.K. and D. N. Raychaudhuri, 1982. One new subgenus, three new species and one new subspecies of aphids (Homoptera) from Sikkim, North east India. *Entomon*, 7, 37–45.

Ahemedullah, M. and M. P. Nayar, 1987. Endemic plants of the Indian Region, *Bot. Surv. India, Calcutta*, 1, 261.

Ali, S. 1981. The Himalaya in the Indian Ornithology. In the Himalaya: Aspects of change, pp. 16–31, edited by J. S. Lall, New Delhi: Oxford University Press, pp.481.

Ananthakrishnan, T. N. 1993. Biological diversity and Conservation Strategies. *Hexapoda*, 5, 5-8.

Annonymous, 1972. Crop pests and how to fight them. Govt. of Maharashtra Publication, pp. 1–242.

Annonymous,1999. Executive Summary on National Forestry Action, Programme India. Ministry of Environment and Forest, Govt. of India, New Delhi.

Balakrishnan, P. and K. M. Alexander, 1984. The endangered wildlife of India and their conservation. *Bull. Ethol. Soc. India.* (*Suppl.*), 3, 181–190.

Basu, A. N. 1961a. Some aphids new to India, with description of new subspecies. *Curr. Sci.*, 30, 390–391.

Basu, A. N. 1961b. Addition to the aphid fauna of West Bengal. *Sci. Cult.*, 27, 436.

Basu, A. N. 1964. New genera and species of aphids from Darjeeling district. India (Homoptera: Aphididae). *J. Linn. Soc.* (*Zool*), 45, 233-243.

Basu, A. N. 1967. One new genus and seven new species of aphids from Darjeeling district, West Bengal (Homoptera: Aphididae). *Bull. Ent.*, 8. 143–157.

Basu, A. N. 1969a. Four new genera and nine new species of aphids (Homoptera) from West Bengal, India. *Oriental Ins.*, 3, 169–186.

Basu, A. N. 1969b. Further records of new and little known aphids (Homoptera) from West Bengal, India. *Oriental Ins.*, 3, 355–371.

Basu, A. N. and S. N. Banerjee, 1958. Aphids of economic plants of West Bengal. *Indian Agric.*, 2, 89–112.

Basu, A. N. and Hille Ris Lambers, D. 1968. One new genus and three new species of Indian aphids (Homoptera: Aphididae). *Ent. Ber. Amst.*, 28, 7–14.

Basu, R. C., Ghosh, A. K. and D. N. Raychaudhuri, 1970. Studies on the aphids (Homoptera: Aphididae) from eastern India. 4. A new genus and records of some sexual forms from Assam. *Proc. Zool. Soc. Calcutta*, 23, 83–91.

Basu, R. C., Ghosh, A. K. and D. N. Raychaudhuri, 1972. A new species of *Eutrichosiphum* and notes on other new records of aphids (Insecta: Homoptera) from NEPA (Arunachal). *Sci. Cult.*, 38, 494–495.

Basu, R. C., Ghosh, A. K. and D. N. Raychaudhuri, 1973. Studies on the aphids (Homoptera: Aphididae) from eastern India. Five new species and thirty new records from Assam. *Proc. zool. Soc., Calcutta*, 26, 89–101.

Basu, R. C., Ghosh, M. R. and D. N. Raychaudhuri, 1974. The genus *Betacallis* with special reference to the species from North East India (Homoptera: Aphididae). *Oriental Ins.*, 8, 233–238.

Behura, B. K. 1963. Aphids of India. A survey of published information. Proc. First Summer School of Zoology, (Simla, 1961), 25–78.

Beason, C. F. C. 1941. Ecology and control of forest insect of India and neighboring countries. Govt. of India (1961 Reprint), pp. 767.

Bhattacharya, D. K., Maity, S. P. and S. Chakrabarti, 1980. Aphids (Homoptera: Aphididae) of North West India. VII Hitherto unknown morphs of three species described from India and Pakistan. *Entomon*, 5(4), 335–339.

Bhoje, P. M. and T. V. Sathe, 2003. Faunistic studies on butterflies from Radhanagari Wildlife Sanctuary, India. *Indian J. Enviorn. and Ecoplan.*, 7(3), 655–658.

Blackman, R. L. and V. F. Eastop, 1985.Aphids on world crops, an identification and information Guide. John Wiley and Sons, N. Y., pp. 1-413.

Blackman, R. L. and V. F. Eastop, 1994. Aphids on the World's trees. C. A. B. International, Wallingford. Bodenheimer, F. S. and F. Swirski 1957. The Aphididae of the Middle East. Weizmann, Jerusalem. pp.

Bonnet, Ch. 1745. Traite d Insectologie. Paris.

Borner, C. 1930. Europe centralis Aphides. De. Blattlause Mitteleuropas: Namen, Synonyme, Wirtsepflanzen, Generationszyken. *Mitt. thuring. bot. Ges.* 3: 1-484.

Buckton, G.B. 1893. A new gall making aphid. *Indian Mus. notes*, 3(1), 71–73.

Buckton, G. B. 1896. Notes on two new species of gall aphids from North Western Himalyan region. *Indian Mus. notes*, 4(2), 50-51.

Buckton, G. B. 1899. The pear tree aphid *Lachnus pyri*. *Indian Mus. Notes*, 4, 274–275.

Burton, T. M. and K. G. Sivaramakrishnan, 1993. Composition of the insect community in the streams of the Sailent Valley national park in southern India. *Trop. Ecol.*, 34, 1-6.

Calow, P. 1978. Life cycles. Champion and Hall, London.

Chakrabarti, S. and D. N. Raychaudhuri, 1974. A new subspecies and some synonymies of aphid species (Homoptera: Aphididae) from India. *Oriental Ins.*, 8, 303–305.

Chakrabarti, S. and D. N. Raychaudhuri, 1978a. (1975). *Callipterine* aphids (Homoptera: Aphididae) of North–East India and Bhutan. *Proc. zool. Soc., Calcutta*, 28, 70–101.

Chakrabarti, S. and D. N. Raychaudhuri, 1978b. New and little known aphids (Homoptera: Aphididae) from Kumaon Himalaya, India. *Entomon*, 3, 95–103.

Chakrabarti, S., Ghosh, A. K. and D. N. Raychaudhuri, 1970.Aphids (Hemiptera) Himachal Pradesh III. *Oriental Ins.*, 4, 447–452.

Chakrabarti, S., Ghosh, A. K. and D. N. Raychaudhuri, 1971. New records of aphids (Insecta Homoptera) from Uttar Pradesh, India. *Sci. Cult.*, 37, 247–248.

Chakrabarti, S., Ghosh, A. K. and D. N. Raychaudhuri, 1972a. On some undescribed morphs and new records of aphids (Homoptera: Insecta) from Kumaon Himalaya, India. *Curr. Sci.*, 41, 70–71.

Chakrabarti, S., Ghosh, A. K. and D. N. Raychaudhuri, 1972b. A new genus, a new species and further records of aphids (Homoptera: Aphididae) from the Kumaon Hills, North West Himalaya, India. *Oriental Ins.*, 6, 387–400.

Champion, H. G. and S. K. Seth, 1968. A revised survey of the forest type of India. Manager of Publication, Delhi, pp. 464.

Chatterjee, N. B., Ghosh, A. K. and D. N. Raychaudhuri, 1961. Aphids of Manipur state. *Indian Agric.*, 5, 87–88.

Chatterjee, P.N. and M.P. Mishra, 1974. Natural insect enemy and plant host complex of forest insect pests of Indian region. *Indian for. Bull. No. 265* (N.S.), Entomology, Controller of Publications, Govt. of India, Delhi, pp. 1-233.

Cherian, P. T. 1996. On hot spots, warm spots and hot specks. *Zoo's Print*, 11(4), 9–11.

Cherian, P. T. 2000. On the status, origin and evolution of hot spots of biodiversity. *Zoo's Print* IS (4), 247–251.

Chelliah, S. and K. Gunathilgaraj, 1993. Biological diversity of insect pests and changing agrotechnology. *Hexapoda*, 5(2), 17-53.

Cottier, W. 1953, Aphids of New Zealand. *Bull. N. Z. Dp. Scient. Ind. Res.*(106),1–132.

Dadhwal, K.S., P. Rai and K.R. Solanki, 2004. Agroforestry for eco-restoration and diaster management. A.P.H. Pub., New Delhi, pp. 1-673.

Das, B. 1918. The Aphidinae of Lahore. *Mem. Indian Mus.*, 6, 135–274.

David, S. K. 1958a. Notes on South Indian Aphids, V. Greenideinae, Thelaxinae and Eriosomatinae. *Indian, J. Ent.*, 20, 124–129.

David, S. K. 1958b. A new genus and three new species of aphids from India. *Indian J. Ent.*, 20, 175–188.

David, S. K. 1969. Some rare aphids in new regions in India. *J. Bombay nat. Hist. Soc.*, 66, 323–326.

David, S. K. 1972. Studies on the systematic of aphid genus *Macrosiphum* (Homoptera: Aphididae). 1–47. (Cyclostyled monograph).

David, S. K. 1975.A taxonomic review of *Macrosiphum* (Homoptera: Aphididae) in India. *Oriental Ins.*, 9, 461–493.

David, S. K. and S. G. Rajasingh, 1969. New records of aphids (Insecta: Homoptera) from Assam, India. *Proc. zool. Soc., Calcutta*, 22, 151–157.

David, S. K., Rajasingh, S. G. and Narayanan, K. 1968. Notes on the taxonomy and other aspects of certain species of aphids in India. *J. Bombay nat. Hist. Soc.*, 65, 508–512.

David, S. K., Narayanan, K. and Rajasingh, S.G. 1970. Five new species of aphids (Homoptera) from northwestern India. *Oriental Ins.*, 4, 413–426.

David, S. K., Rajasingh, S. G. and K. Narayanan, 1970. New aphids (Homoptera) from Darjeeling in northeastern India. *Oriental Ins.*, 4, 407–412.

David, S. K., Narayanan, K. and S.G. Rajasingh, 1971. Records of aphids in new region in India. *Madras agric. J.*, 58, 372–374.

David, S. K., Rajasingh, S. G. and S. Narayanan, 1972. New genus, new species and new morphs of aphids (Homoptera) from India. *Oriental Ins.*, 6, 35–44.

Dixon, A. F. G. 1958. Escape responses shown by certain aphids to the presence of coccinelid *Adelia decempunctata* (L). *Trans. R. Ent. Soc. Lond.*, 10, 319–334.

Dixon, A. F. G. 1985. Aphid ecology. Blackie Glasgow, pp.1–157.

Dreeze, J. and A. Sen, 1995. *Economic development and social opportunity*, New Delhi. Oxford University Press, pp. 291-292.

Eastop, V. F. 1958. A study of the Aphididae (Homoptera) of East Africa. H. M. S. O., London.

Eastop, V. F. 1959. New African *Sitobion* (Aphididae: Homoptera). *Entomologist*, 92, 96–105.

Eastop, V. F. 1961. A study of the Aphididae (Homoptera) of West Africa. H. M. S. O., London. pp.

Eastop, V. F. 1966. A taxonomic study of Australian Aphidoidea (Homoptera). *Austr.J. Zool.*, 14, 399–592.

Eastop, V. F. 1977. Worldwide importance of aphids as virus vectors. Eds K. F. Haris and Maramorosh, Academic Press, London, pp. 3-62.

Eastop, V. F. and H. F. Van Emden, 1972. "The Insect Material" pp.1–45, in Aphid technology, ed.H. F. Van Emden. Academic Press London and New York. 1–344.

Essig, E. O. and S. I. Kuwana, 1918. Some Japanese Aphididae. *Proc. Calif. Acad. Sci,*. 8(3), 35–112.

Fotedar, M. R. and A. P. Kapur, 1943. First record of the sexual forms and oviparous reproduction of woolly aphid. *Eriosoma lanigerum* Hausm, from Kashmir. *Curr. Sci.*, 12, 84–85.

Ganguli, R. N. and M. R. Ghosh, 1965. A note on the aphids of economic importance in Tripura. *Sci. Cult.*, 31, 541–542.

Gaonkar, H. 1996. The butterflies of Western Ghats, India and Srilanka. A biodiversity assessment on the ecological history of Western Ghats. Abst., pp. 99.

George, C. J. 1927. South Indian Aphididae. *Soc. Bengal (U. S.)*, 25, 1-12.

Ghosh, A. K. 1969. New aphids (Insecta: Homoptera) from North Bengal, India. *Proc. zool. Soc., Calcutta*, 22, 121–127.

Ghosh, A. K. 1973. Taxonomic notes on some species of Indian Homoptera. *Oriental Ins.*, 7, 347–350.

Ghosh, A. K. 1974a. New species and new records of aphids (Homoptera:Aphididae) from Northeast India. *Oriental Ins.*, 8, 161–175.

Ghosh, A. K. 1974b Some new or little known species of aphids (Homoptera: Aphididae) from India. *Oriental Ins.*, 8, 425–535.

Ghosh, A. K. 1974c. A list of aphids (Homoptera: Aphididae) from India and adjacent countries. *J. Bombay nat. Hist. Soc.* 71, 210–225.

Ghosh, A. K. 1980. General Introduction and subfamily Chaitophorinae, Part-I. *Zool. Sur. India*, pp. 1–122.

Ghosh, A. K. 1982. Fauna of India–Homoptera–Aphidoidea, Subfamily–Lachinae, Part–II. *Zool. Sur. India*, pp. 1–167.

Ghosh, A. K. 1984. Fauna of India–Homoptera–Aphidoidea, Subfamily–Pemphiginae, Part–III. *Zool. Sur. India*, pp. 1–424.

Ghosh, A. K. 1988. Fauna of India and adjacent countries–Homoptera-Aphidoidea.Subfamily-Phloemyzinae, Anoecinae, Hormaphidinae, Part-IV. *Zool. Sur. India*, pp. 1–322.

Ghosh, A. K. and B. K. Agawala, 1993. Fauna of India–Homoptera–Greenideinae, Part–VI. *Zool. Sur. India*, pp. 1–321

Ghosh, A. K. and D. N. Raychaudhuri, 1959. Aphids of Calcutta and its suburbs (West Bengal). *J. Bombay nat. Hist. Soc.*, 56, 661–664.

Ghosh, A. K. and Raychaudhuri, D.N. 1962a. A preliminary account of bionomics and taxonomy of aphids from Assam.II. *J. Asiat. Soc., Calcutta*, 4, 101–113.

Ghosh, A. K. and D.N. Raychaudhuri, 1962b. A preliminary account of bionomics and taxonomy of aphids from Assam. *J. Asiat. Soc., Calcutta*, 59, 238–253.

Ghosh, A. K. and D. N. Raychaudhuri, 1968a. Two new records of aphids (Homoptera: Aphididae) from India. *Bull. Ent.*, 9, 147–148.

Ghosh, A. K. and D. N. Raychaudhuri, 1968b. Aphids (Insecta: Homoptera) of Sikkim. *Proc. zool. Soc., Calcutta*, 21, 179–195.

Ghosh, A. K. and D. N. Raychaudhuri, 1968c. New aphids from North East India (Homoptera). *Ann. ent. Soc. Am.*, 61, 752–755.

Ghosh, A. K. and D. N. Raychaudhuri, 1969. A note of *Chaetomyzus*, *Ericophilum* and *Neoacyrthosiphon* including a new subgenus *Pseudoacrythosiphon* (Homoptera) from India. *Oriental Ins.*, 3, 93–96.

Ghosh, A. K. and D. N. Raychaudhuri, 1971. Further records of aphids (Homoptera: Insecta) from NEFA. *Sci. Cult*, 37, 483–484.

Ghosh, A. K. and D. N. Raychaudhuri, 1972a. Studies on the aphids (Hemiptera: Aphididae) from eastern India. 16. Description of new subgenus and seven new species and notes on two little known species. *Proc. zool. Soc., Calcutta*, 25, 93–107.

Ghosh, A. K. and D. N. Raychaudhuri, 1972b. Studies on the aphids (Homoptera: Aphididae) from eastern India XIII. New species and new records from Assam. *Oriental Ins.*, 6, 371–386.

Ghosh, A. K. and Raychaudhuri, D. N. 1973. Studies on the aphids (Homoptera: Aphididae) from Eastern India. XV. A study of *Nipponaphis* Pergande and related genera with descriptions of new genus and eight new species from eastern India. *Kontyu*, 41, 148–165.

Ghosh, A. K. and F. W. Quednau, 1988. Fauna of India Homoptera, Aphidoidea. Subfamily-Drephanosiphinae, Part–V. *Zool. Sur. India*, pp. 1–331.

Ghosh, A. K., Basu, R. C. and D. N. Raychaudhuri, 1969. A new genus and seven new species of aphids (Homoptera) from India. *Oriental Ins.*, 3, 245–254.

Ghosh, A. K., Basu, R. C. and D. N. Raychaudhuri, 1970. Studies on aphids (Homoptera: Aphididae) from eastern India. New species and new records from Assam. *Oriental Ins.*, 4, 65–76.

Ghosh, A. K., Basu, R. C. and D. N. Raychaudhuri, 1971. On a collection of aphids (Homoptera: Aphididae) from Bhutan with descriptions of two new species. *Kontyu*, 39, 120–125.

Ghosh, A. K., Banerjee, H. and D. N. Raychaudhuri, 1971a. Aphids (Homoptera: Insecta) collected from Kaneng district, NEFA. *Gaveshana*, 3, 10.

Ghosh, A. K., Banerjee, H. and D. N. Raychaudhuri, 1971b. Studies on the aphids (Homoptera: Aphididae) from eastern India. V. New species, new subspecies, new records and sexuality of some species of aphids from NEFA and Assam. *Oriental Ins.*, 5, 103–110.

Ghosh, A. K., Banerjee, H. and D. N. Raychaudhuri, 1971c. Aphids (Homoptera: Insecta) species collected from Manipur and Nagaland. *Gaveshana*, 3, 60–61.

Ghosh, A. K., Banerjee, H. and D. N. Raychaudhuri, 1971d. Studies on the aphids (Homoptera: Aphididae) from eastern India 10. A new species and further new records from Sikkim. *Proc. zool. Soc., Calcutta*, 24, 99-111.

Ghosh, A. K., Ghosh, M. R. and D. N. Raychaudhuri, 1970. Studies on the aphids (Homoptera: Aphididae) from eastern India. II. Some new species and new records from North Bengal. *Oriental Ins.*, 4, 193–203.

Ghosh, A. K., Ghosh, M. R. and D. N. Raychaudhuri, 1971a. Studies on the aphids (Homoptera: Aphididae) from eastern India. VII. New species and new records from West Bengal. *Oriental Ins.*, 5, 209–222.

Ghosh, A. K., Ghosh, M. R. and Raychaudhuri, D. N. 1971b. Studies on aphids (Homoptera: Aphididae) from eastern India. IX. One new genus, five new species, a new subspecies and further new records from Darjeeling district, West Bengal. *Oriental Ins.*, 5, 328–336.

Ghosh, A. K., Ghosh, M. R. and D. N. Raychaudhuri, 1972. Studies on aphids (Homoptera: Aphididae) from eastern India. XI. Descriptions of hitherto unknown or newly recorded sexual morphs of some species from West Bengal. *Oriental Ins.*, 6, 333–342.

Ghosh, L. K. 1969. On a collection of aphids (Homoptera: Aphididae) from Uttar Pradesh, India. *Sci. Cult.*, 35, 493.

Ghosh, L. K. 1970a. A new species of Megoura.

Ghosh, L. K. 1970b. Aphids (Homoptera: Insecta) of NEFA. India. *Sci. Cult.*, 36, 562–563.

Ghosh, L. K. 1972a. Notes on collection of aphids from Bhutan with description of a new species (Homoptera: Aphididae). *Kontyu*, 40, 72–76.

Ghosh, L. K. 1972b. A new species and subspecies of aphids (Homoptera: Aphididae) from India. *Oriental Ins.*, 6, 299–304.

Ghosh, M. R., Ghosh, A. K. and D. N. Raychaudhuri, 1970. Studies on aphids (Homoptera: Aphididae) from eastern India. III. New genus and new species and new records from North Bengal and Sikkim. *Oriental Ins.*, 4, 377–393.

Ghosh, M. R., Ghosh, A. K. and D. N. Raychaudhuri, 1971a. Studies on the aphids (Homoptera: Aphididae) from eastern India. 6. New records from Sikkim. *Proc. zool. Soc., Calcutta*, 24, 47–51.

Ghosh, M. R., Ghosh, A. K. and D. N. Raychaudhuri, 1971b. Studies on the aphids (Homoptera: Aphididae) from eastern India. 12. New records and undescribed sexual forms of two species of Sikkim. *Proc. zool. Soc., Calcutta*, 24, 163–168.

Ghosh, M. R., Basu, R. C. and D. N. Raychaudhuri, 1976. Studies on the aphids (Homoptera: Aphididae) from eastern India. XXVIII. Acutosiphon, Akkaia and Vesiculaphis. *Oriental Ins.*, 10, 267–276.

Goot, P. Van Der. 1916. On some undescribed aphids from the collection of the Indian Museum. *Rec. Indian Mus.*, 12(1), 2–3.

Goot, P. Van. Der. 1917. Notes on some Indian Aphids. *Rec. Indian Mus.*, 13, 175–183.

Gulamullah, C., 1940. Studies on Indian Aphididae I. The aphid fauna of Delhi. *Indian J. Ent.*, 2, 13-25.

Gupta, R.K. 2004. Advancements in insect biodiversity. *Agrobios* (India), Jodhpur, pp. 1-324.

Heie, O. E. 1967. Studies on fossil aphids (Homoptera: Aphidoidea). *Spolia Zool. Mus. Hauniensis*, 26, 7–274.

Heinze, K. 1961. Systematic mitteleropaischen Myzinae. *Beitr. Ent.*, 11, 24–96.

Hille Ris Lambers D. 1938. Contribution on the monographs of Aphididae of Europe, I. *Temminckia*, 3, 1–44.

Hille, Ris Lambers D. 1939, Contribution on a monograph of the Aphididae of Europe. II. *Temminckia*, 3, 1–44.

Hille, Ris Lambers D. 1947. Contribution on the monographs of Aphididae of Europe, III. *Temminckia*, 7, 179–319.

Hille, Ris Lambers D. 1949. Contribution on the monographs of Aphididae of Europe, IV. *Temminckia*, 8, 182–323.

Hille, Ris Lambers D. 1953a. Contribution on the monographs of Aphididae of Europe, V. *Temminckia*, 9, 1–176.

Hille, Ris Lambers, D. and Basu, A. N. 1966b. Some new or little known genera, subgenera, and species from India (Homoptera: Aphididae). *Ent. Bericht. Amst*, 26, 27–36.

Huxley, T. H. 1858. On the agamic reproduction and morphology of Aphids, Part I. *Trans. Linn. Soc.*, 22, 193–219.

Jadhav, B. V. and Sathe, T. V. 2006a. Biodiversity of aphids (Order: Hemiptera) from Satara district of Western Ghats. *Indian J. Enviorn. and Ecoplan.*, 12(1), 237–240.

Jadhav, B. V. and Sathe, T. V. 2006b. Biodiversity of aphids (Order: Hemiptera) from Poona district of Western Ghats. *J. Adv. Zool.*, 27(1), 43–45.

Jalan, B. 1996. India's Economic Policy: Preparing for the 21st Century, New Delhi. Penguin Books. pp. 217.

Joseph, K. J. 1984. Insect life and ecodevelopment of the Western Ghats. *Proc. Semi. on Ecodevelopment of Western Ghats*, 1986, pp.84–89.

Joseph, T. M. 2004. Biodiversity conservation in the Western Ghats. *Biodiversity and Environment*, 17, 221–227.

Krishnamurthy, B. 1928. Aphididae of Mysore. *J. Bombay nat. Hist. Soc.*, 33, 211–215. Khan, M. S. 1999. Current problems of Indian economy. Asia Publishing House, New Delhi, 282.

Krebs, C. J. 1989. Ecological Methodology. Harper and Raw Publishers, New York.

Khoshoo, T. N. 1995b. Census biodiversity takes ahead. *Curr. Sci.*, 69, 14–17.

Kumar, R. and Lavinge, 1970. A new genus and new species of *Callipterine* from India (Homoptera: Aphididae). *Pan. Pacif. Ent.*, 46, 120–124.

Lal K. B. and Z. A. Sddiqui. 1952. Biology of peach leaf curling aphid on the Kumaun Hills. *Indian J. Ent.* 14, 191–196.

Larson, T.B. 1987. The butterflies of the Nilgiri mountains of South India (Lepidoptera: Rhopalopcera). *J. Bombay nat. Soc.*, 84(1), 26-54; 84(12), 291-316; 84(3), 560-584.

Maity, S. P. 1980. Aphids of North West India with special reference to Garhwal, Himalaya. Ph. D. Thesis, University of Kalyani, 1–441.

McNeely, J. A., Miller, K. R., Reid, W. V., Mittermeier, R. A. and T. B. Werner, 1990. Conserving the worlds' Biodiversity. Washington, DC: World Resource Institute, pp193.

Menon, Ramdas M. and Pawar, J. C. 1968. Taxonomic studies of Indian Aphididae I. Additions in the aphid fauna of Delhi. *Proc. 45th Indian Sci. Cong.*, 3, 348.

Mittermeier, R. A., N. Myers, J. B. Thomson. G. A. B. da Fonesca and D. Oliveri, 1988. Biodiversity hot spots and major tropical wilderness areas: Approaches to setting conservation priorities. *Conservation Biology*, 12(3), 516–520.

Miyazaki, M. 1971. A revision of the tribe Macrosiphini of Japan (Homoptera: Aphididae, Aphidinae). *Inst. Matsum*, 34, 1-247.

Mordvilko, A.K. 1914. *Raune Russie. Insects Hemiptera*, 1(1), 1-236.

Mordvilko, A.K. 1919. Faune Russie, *Insect Hemiptera*, 1(2), 237-508.

Mordivilko, A. 1928. The evolution of cycles and the origin of heteroecy (migration) in plant lice. *Ann. Mag. nat. Hist.*, 2(10), 570–582.

Mordivilko, A. K. 1935. Die Blattlause mit unvollstandigen Generations zyklus undhihre Entstehung. *Ergebn. Zool.* 8, 36–328.

Myers, N. 1988. The Environmentalists, 8, 187–208.

Myers, N. 1990. The Environmentalists, 10, 243–256.

Mondal, P. K. 1977. Aphids of Sikkim. Ph. D. Thesis, University of Calcutta., 1–353.

Nagendran, N.A. and M.S. Smija, 2004. A seminal report on diversity of higher taxa of aquatic insects in three hills streams of Western Ghats, South India. A.P.H. Pub. Corp. New Delhi, pp. 1-659.

Nair, K.S.S., Mathew, G. and Shivarajan M. 1973. Occurrence of bagworm *Pterowea plagiophleps* Hampson (Lepidoptera: Psychidae) as a pest of the tree *Albizia* falcataria in Kerala, India. *Entomon.*, 6, 179-180.

Nair, K.S.S. and G. Mathew, 1993. Diversity of insects in Indian forests. The State of our knowledge. *Hexapoda*, 5(2), 71-78.

Pal, P. K. 1975. Studies on Aphids of Graminaceous plants in North Eastern Himalayas with special reference to those infesting the subaerial parts. Ph. D. Thesis, University of Calcutta. 1–222.

Palmer, M. A. 1952. Aphids of Rocky Mountains regions, 5: 1–452. The Thomas Say Foundation.

Pandharbale, A. R. and T. V. Sathe, 2001. On a new species of the genus *Syntomis* (Syntomidae: Lepidoptera) from the environment of Western Ghats (Satara District). *Indian J. Environ. and Ecoplan.*, 5(2), 601–602.

Paroda, R. B. 1993. Crop science research for sustainable agriculture in India. *Proc. Nat. Acad. Sci. India* (*B*), 63, 97–114.

Pergande, 1906. Related genera with descriptions of new genus and eight new species from eastern India. *Kontyu*, 41, 148–165.

Quednau, F.W. and S. Chakrabarti, 1976. A new genus and three new species of aphids (Homoptera: Aphididae) on alder and birch in India. *Can. Ent*, 117, 227–232.

Rathakrishnan, L. 2003, India's environmental problems and issues. In Environment and its challenges. (Ed. Arvind Kumar). 13, 34-38.

Raychaudhuri, D. N. 1956. Revision of *Greenidea* and its related genera. *Zool. Verh. Leiden*. 31, 1–106.

Raychaudhuri, D. N. (ed) 1973. Taxonomy of the aphids of the Eastern Himalayas. PL 480 Technical Report (cyclostyled), 1–107.

Raychaudhuri, D. N. 1974. A note on the status of the genus *Glyphina* Koch and Kurisakia Takahashi. *Sci. Cult.*, 40, 107–110.

Raychaudhuri, D. N. 1980. Aphids of North–East India and Bhutan. *Zool. Sur. India*. pp. 1–459.

Raychaudhuri, D. N. and C. Banerjee, 1974. A study of the genus *Melanaphis* (Homoptera: Aphididae) with descriptions of new taxa from India. *Oriental Ins.*, 8, 365–389.

Raychaudhuri, D. N. and M. Chatterjee, 1977. Studies on the aphids (Homoptera: Aphididae) from eastern India. XXXII. Two new species and new morph of Greeideinae from West Bengal. *Oriental Ins.*, 11, 263–267.

Raychaudhuri, D. N. and M. Chatterjee, 1980. Subfamily Greeneideinae 314–358. In: "Aphids of North-East India and Bhutan"(D.N. Raychaudhuri, ed.) *The zool. Soc., Calcutta*, vi+ 521 pp.

Raychaudhuri, D. N. and A. K. Ghosh, 1958. A note on aphids of Calcutta and suburbs, with special reference to the newly records host plant families for some of the species. *Curr. Sci.*, 27, 402.

Raychaudhuri, D. N. and P. K. Pal, 1974. Studies on the aphids (Homoptera: Aphididae) from eastern India. XXII. Description of a new genus *Subtakecallis* and a new species *Subtakecallis brevisetosus*. *Oriental Ins.*, 8, 95–98.

Raychaudhuri, D. N, Ghosh, M. R., Banerjee, M and A. K. Ghosh, 1973. Studies on the aphids (Homoptera: Aphididae) from eastern India-14. One new genus, two new subgenera and fifteen new species of Greenideini. *Kontyu*, 41, 53–73.

Rojers, W. A. 1991. In Ecology and sustainable development. (Ed. Gopal), pp. 81–83.

Rodger, W. A. and H. S. Panwar, 1988. Planning of wildlife protected area network in India.-1 and 2, Dehra Dun: Wildlife Institute of India.

Roonwal, M.L., G.D. Bhasin and G.D. Pant, 1950. A systematic catalogue of the main identified entomological collection at the Forest Research Insitute, Dehra Dun. Parts 1-3. *Indian for.* 76(11), 498-505.

Sathe, T. V. 1992. Fauna of aphids on plants of economic importance found in Western Maharashtra, India. *J. Curr. Biosci.*, 9(1), 27–31.

Sathe, T. V. 2004a. Biology and behaviour of Coccinellid beetles, in Indian insect predators in biological control. Ed. K. Sahayaraj, 8, 177–198.

Sathe, T.V. 2004b. Vermiculture and Organic Farming, D.P.H., New Delhi, pp. 1-122.

Sathe, T.V. 2008. A Textbook of Forest Entomology, D.P.H., New Delhi.

Sathe, T.V. and P. M. Bhoje, 2000. Biological control of insect pests. D.P.H., New Delhi pp. 1-121.

Sathe, T.V. and Y.A. Bhosale, 2001. Indian Pest Predators, DPH, New Delhi, pp. 1-167.

Sathe, T. V. and G. S. Margaj, 2001. Cotton pests and Biocontrol agents. DPH, New Delhi. pp. 1-161.

Sathe, T. V. and M. K. Mulla, 1995. Insect pests of Mulberry from Amboli. *Oikoassay*, 12(1–2), 9.

Sathe, T. V. and A. R. Pandharbale, 1999. Hawk moth (Sphingidae: Lepidoptera) diversity in Western Maharashtra including Ghats. *Geobios*, 18, 77–82.

Sathe, T. V. and A. R. Pandharbale, 2005. Biodiversity of Moths (Order–Lepidoptera) from Western Ghats of Satara district, India. *Bull. Biol. Sci.* 1, 81–88.

Sathe, T.V. and Patil V.J. 2003. Insect predators and pest management, DPH, New Delhi pp. 1-165.

Sathe, T.V., Inamdar, S. A. and M. V. Santhakumar, 1986–1987. Fauna of Butterflies from Western Maharashtra and Western Ghats (Part of Maharashtra only), India. *J. Shivaji Uni.* (*Science*), 23, 391–398.

Sathe, T.V. and K. P. Shinde, 2006. Diversity of butterflies from Western Ghats (Kolhapur district). *J. Nat. Con.*, 81 (1), 181–184.

Sathe, T.V. and K. P. Shinde, 2006a. On a new species of the genus *Agriochemis* Selys (Order: Odonata) from India. *J. Curr. Sci.*, 9(2): 799–802.

Sathe, T.V. and K. P. Shinde, 2006b. On a new species of the genus *Rhiothemis* Wagen (Order: Odonata) from India. *J. Nat. Con.*, 18(2): 421–424.

Schoutenden, H. 1905. Notes on Ceylonese Aphids. *Sopolia Zeylan*, 2: 181–183.

Shaposhnikov, G. Kh. 1964. Suborder Aphidinea. In Bei Beinko G. Y. (ed): Key to the insects of Europhian parts of U.S.S.R., 1.: 489–616. Zool. Inst. Acad. Sci. USSR, Leningrad.

Shaposhnikov, G. Kh. 1977. The trends of evolution. *Zhurnal Obshchei Biologii*, 38, 649–655.

Shaposhnikov, G. Kh. 1981. Populations and species in aphids and the need for a universal species concept. Special Publication, Research Branch, Agriculture, Canada.

Shinde, K. P. 2006. Biodiversity of dragonflies (Order–Odonata) from Koyana dam and around area. Ph. D. Thesis, Shivaji University, Kolhapur, pp.1–275.

Singh, A. and A. N. Sadhu, 1986. Agricultural problems in India. Jammu University, Jammu, pp. 386.

Sivaramkrishnan, K. G, Sridhar, S. and Venkataraman, K. 1990. Habitat, microdistruction, life cycle patterns and tropic relationship of mayflies of Cardamam hills, Western Ghats. *Hexapoda*, 11, 59–64.

Sivaramakrishnan, K. G., Venkataraman, K., Moorthy, R. K., Utkarsh, G. and K. A. Subramanian, 2000. Aquatic insect diversity and ubiquity in the streams of the Western Ghats, India. *J. Indian Inst. Sci.*, 80, 537–552.

Smith, C. F. 1972. Bibliography of Aphidinae of the world. *Tech. Bull. N. Card. Agric. Exp. Sin.* (216), 1–717.

Stebbing, E.P. 1977. Indian forest insects of economic importance–Coleoptera. J.K.J. Brothers, Bhopal, pp. 1-648.

Stary, P. and A. K. Ghosh, 1978. Further records of aphid parasitoids (Hymenoptera: Aphididae) from Megalaya. India. *Oriental Ins.*, 12(1), 77-80.

Stroyan, H. L. G. 1954. A new sub genus and species of *Myzus* Passerini, 1860 (Hemiptera: Aphididae). *Proc. R. ent. Soc. Lond.*(B), 23, 10-16.

Stroyan, H. L. G. 1964. A new subspecies of Acyrthosiphon Mordvilko, 1914 (Homoptera: Aphidoidea) from Rubus ellipticus in North India. *Proc. R. ent. Soc. Lond.*, 36, 186.

Szelegiewich, H. 1969. *Bibliogafia–Mszyce Szkodniki Roslin wektory choros, wirusowych I produceni spadzi*, Pantwowe Wydamnictwo Ronicze Lesne, Warszawa.

Takahashi, R. 1921. Aphididae of Formosa, 1. Dep. Agric. Govt. Res. Inst., Formosa.

Takahashi, R. 1924. Aphididae of Formosa, 3. Dep. Agric. Govt. Res. Inst., Formosa.

Takahashi, R. 1925. Aphididae of Formosa, 4. Dep. Agric. Govt. Res. Inst., Formosa.

Takahashi, R. 1929. Notes on some Formosan Aphididae, 2. *Trans. Nat. Hist. Soc. Formosa*, 19, 247–259.

Takahashi, R. 1931. Aphididae of Formosa, Part VI. *Rep. Dep. Agric. Govt. Res. Inst. Formosa*, 53, 1–127.

Takahashi, R. 1958. On the aphid of Ceratovacuna in Japan (Aphididae–Homoptera). *Ann. ent. Soc. Am.* 43, 587–607.

Takahashi, R. 1961. A new genus and four new species of Aphidinae (Homoptera–Aphididae). *Insect Matsum*,24, 104–111.

Takahashi, R. 1965. *Myzus* of Japan (Aphididae). *Mushi.*, 38, 43-78.

Theobald, F.V. 1915. African Aphididae 2. *Bull. ent. Res.* 6, 103-153.

Verma, K. D. 1965. Additions to the aphid fauna of North–West India. *Sci. Cult.*, 31, 389.

Verma, K. D. 1969. A new genus, some new species and some new records of aphids from N. W. India. *Sci. Cult.*, 35, 28-29.

Vohra, B. B. 1980. A charter for land, Sardar Patel Memorial lecture.

Way, M. J. and C. J. Banks, 1967. Intraspecific mechanism in relation to the natural regulation of number of *Aphis fabae* Scop. *Ann. Appl. Biol.*, 59, 189–205.

Wyner, Blyth, M. A. 1957. Butterflies of Indian Region. *Bom. nat. Hist. Soc., Bombay*, 1–523.

Yadav, S. R. and M. B. Bachulkar, 1995. New plant record from Satara district Part-I. Botanical Reporter journal Vol. 14(1 and 2), 1-6.

Yadav, S. R. and M. M. Sardesai, 2002. Flora of Kolhapur District. Pub. *Shivaji Uni. Kolhapur, India*. pp. 1–680.

Zoological Survey of India, 1983. Threatened Animals of India, Calcutta *Zoological Survey of India*. pp. 307.

Index

A

Abaca 29
Acacia catechu 58, 110, 138
Acrythosiphon pisum Harris 97
Acutospiohon (Basu) 51
Acyrthosiphon 90
Acyrthosiphon (Mordvilko) 51, 89
Acyrthosiphon calatropi sp.nov. 90, 91, 98
Acyrthosiphon picum Harris 93
Acyrthosiphon vatani sp.nov. 94, 95, 98
Adelia decempunctata (L) 192
Akkaia (Takahashi) 51
Aleurodaphis (Goot) 53
Allotrichosiphum (Takahashi) 53
Ambadi, *Hibiscus* 23
Amphicercidus (Destland) 51
Amphorophora (Backton) 51, 90
Amphorophora ampulata Buckton 90
Anacardium occidentale Linn. 83
Anomalosiphum (Takahashi) 37, 53, 187
Anona 31
Aphids 2
Aphis acaciae sp.nov. 55, 56, 62
Aphis adusta 12
Aphis aurantii Boyer de Fonscolombe 73
Aphis bambussae Fullaway 67
Aphis brassicae Linnaeus 99
Aphis cerasi Fabricius 122
Aphis craccivora 3, 17, 21, 59, 60, 62
Aphis diharda Walker 90
Aphis erysimi Kaltenbach 105

Aphis gossypii sp.nov. 22, 24, 58, 63, 64, 72
Aphis jaswandi sp.nov. 62
Aphis kaneri sp.nov. 62
Aphis Linnaeus 51
Aphis malvae 23, 24
Aphis medicaginis 20
Aphis nymphaceae Linnaeus 79
Aphis pisi Kaltenbach 89
Aphis rosae Linnaeus 138
Aphis sambuci 54
Aphis urticae Schrank 90
Apple 8
Arachis hypogea 3, 63
Arthromyzus (Borner) 51
Artocarpus integrifolia 167, 168
Astegopteryx (Karsh) 53
Aulacorthum (Mordvilko) 51, 146
Azardirecta indica 77

B

Bajra 11
Bamboo sp. 177
Banana 24, 29
Barley 8, 12
Bean 8, 19, 21, 23, 58
Brachimyzus (Basu) 51
Brachycaudus (Van der Goot) 51
Brachycaudus helichrysi 27
Brassica botrytis 110
Brassica capitata 110, 138
Brevicoryne (Van der Goot) 51, 99, 100
Brevicoryne brassicae Linnaeus 25, 103
Brevicoryne capsici sp.nov. 100, 101, 104
Brevitrichosiphon (Raychaudhuri) 53
Brumoides sp. 94, 110, 162
Brussels 25
Bryophylum pinnatum 138

C

Cabbage 8, 25
Cajanus cajan Mill 63, 66, 87, 110
Caladium 29
Calatrophis gigantia 93
Callipterine 197
Capitophorus (Van der Goot) 51
Capsicum annum Linn. 103
Carthamus tinctorius 3
Casimira Eastop 51
Cassic glauca 22
Cauliflower 8, 25
Cavariclla (Guercio) 51
Cerataphis (Lichtenstein) 53
Ceratoglyphina (Goot) 54
Ceratovacuna (Zehntner) 54
Ceratovacuna bambusi sp.nov. 174, 175, 178
Ceratovacuna lanigera sp.nov 15, 180, 181, 183
Ceratovacuna lanigera Zehntner 173, 179
Ceratovacuna spinulosa Ghosh and Raychaudhuri 177
Cereals 8
Cervaphis (Goot) 53
Chaetomyzus (Ghosh and Raychaudhuri) 51, 194

Chaitoregma (Lambers) 54
Chilies 8, 24
Chinarose 22
Cineraria 27
Citrus 31
Coccinella septum-punctata 19
Coccinella sp. 58, 63, 66, 89, 99, 145, 169, 185
Coccinellid beetle 121
Coffee 31
Colarodoa (Wilson) 51
Conobathra aphidovora 16, 185
Corn 8
Cotton 8, 22
Cow pea 21, 58
Cruciferous 19
Cryptosiphon (Oestlund) 51
Cryptosiphum (Buckton) 51
Cryso-perla carnea 17
Crysopa scelestes 19
Crysopa sp. 110
Crysoperla carnea 99
Cucumber 27
Cucurbits 23, 24
Curcumba 29
Cyanodon dactyiedon 15

D

Dactynotus (Rafinsque) 51
Dactynotus carthami 20
Dactynotus compositae 3
Dancus carrota 138
Delphacids 2
Diaretialla rapae 19, 25
Diphorodon (Borner) 51
Dolichos lablab 58
Dysaphis (Borner) 51

E

Eomyzus (Takahashi) 51
Ericophilum 194
Eriosoma lanigerum Hausm 25, 192
Eucalyptus sp. 66
Eudema angulatum 138
Eumyzus (Shinji) 51
Euthoracaphis (Takahashi) 54
Eutrichosiphum (Essia and Kuwana) 53, 151, 153, 188
Eutrichosiphum neoparatricho-siphum radhanagari sp. 157, 158, 160, 161
Eutrichosiphum paratricho-siphum alnicola Basu 152, 153, 157
Eutrichosiphum paratricho-siphum pisdiumi sp.nov. 153, 154, 157, 161

F

Ficus 31

G

Geocia lucifera 14
Gliricidia maculata Syn. 122
Glycine soja 3
Glyphinaphis (Goot) 54
Gossypium arboreum 66
Greenidea (Schoutendon) 53
Greenidea Greenidea bhima-shankari sp.nov. 170, 172
Greenidea Greenidea fhanasae sp.nov. 164, 165, 168
Greenidea photiniphaga, Ray-chaudhuri and Ghosh 169

Greenidea Schoutenden 162, 163
Greenideoidea (Goot) 53

H

Hayhurstia (Gluercio) 51
Helianthus annus 63
Heliconia 29
Hillerislambersia (Basu) 51
Holotrichosiphum (Raychaudhuri) 53
Hover fly 66, 105, 157, 169, 185
Hyadaphis (Kirkarldy) 51, 100, 106
Hyalomyzus (Richards) 51
Hyalopterus Koch 51
Hyperomyzus (Borner) 51
Hysteroneura Davis 51

I

Illeis sp. 132, 173
Impatientinum (Mordvilko) 51
Indaphis (Basu) 51
Indomipponaphis (Ghosh) 54
Indomyzus (Ghosh) 51, 116, 117
Indomyzus glaricae sp.nov. 117, 118, 121
Indomyzus sensoriatus Ghosh, Ghosh and Raychaudhuri 116, 122
Indumasonaphis (Raychaudhuri) 51
Ischiodon scutellaris 18
Ishiodon javana 18, 63

J

Jacksonia (Theobald) 51
Jassids 2
Jowar 11, 14

K

Knol-knol 8
Kurisakia (Takahashi) 54

L

Lacewing 13, 17, 20, 66, 73, 89, 99, 116, 122, 138, 145, 157, 169, 185
Lady bird beetles 12, 13, 20, 22, 32, 116, 122, 162, 179
Lady's finger 8
Leucaena 110
Liosomaphis (Walker) 51
Lipaphis (Mordvilko) 51, 100, 105, 106
Lipaphis erysimi 18, 107, 115, 106
Lipolexis gracilus 19
Lycopersicon esculeatum 66

M

Macromyzella (Ghosh) 51
Macromyzus (Takahashi) 51
Macrosiphoniella (Guercio) 51
Macrosiphum 139
Macrosiphum rosae 32
Macrosiphum (Passirini) 51, 138
Macrosiphum granarium Kby 12
Macrosiphum koyenensis sp.nov. 140, 141, 142, 150
Macrosiphum miscanthi (Takahashi) 13
Macrosiphum pisi 21
Macrosiphum pseudoluteum 139
Macrosiphum rosaeiformis 32
Macrosiphum spinotibium Ghosh, Ghosh and Raychaudhary 144

Maize 11
Mangifera indica 63, 66, 77
Mango 31
Masonaphis (Lamber) 51
Matsmurja (Schumacher) 51
Melanaphis 36
Melanaphis sacchari Zehntner 13, 71
Melanaphis Van der Goot 51, 67
Melanaphis vulgari sp.nov. 68, 69, 72
Melons 8
Membracids 2
Menochilus sexmaculatus 11, 18, 19
Menochilus sp. 63, 66, 73, 84, 89, 94, 105, 128, 138, 151, 157, 169, 185
Metanipponaphis (Takahashi) 54
Metapolophium Mordvilko 51, 90
Micruspis sp. 173
Microlophium Mordvilko 90
Micromyzodium (David) 51
Micromyzus van der Goot 51, 117
Microphis sp 145
Millets 12
Miscanthus sinensis 15
Mollitrichosiphum (Suenaga) 53
Mung 21
Musa paradisiaca Linn. 24, 114
Mustard 8, 19, 23, 25
Myzakkaia (Basu) 51
Myzus (Passirini) 51, 122
Myzus abtusirostris David, Narayanan and Rajasingh 131
Myzus ajarai sp.nov. 124, 125, 137
Myzus persicae sp.nov. 19, 23, 137, 132, 133, 134
Myzus terminalae sp.nov. 128, 129, 137

N

Nectaorosiphon 123
Neoacyrthosiphon (Tao) 51, 194
Neoceratovacuna (Ghosh) 54
Neohylamyzus (Basu) 51
Neomacrosiphum 139
Neomegaurophis (Ghosh) 51
Neomyzus 146
Neomyzus (Goot) 51
Neomyzus circumflexus Buckton 151
Neomyzus koyani sp.nov. 146, 147, 150
Neoparatrichosiphum 153
Neothoracaphis (Matsmura) 54
Neotoxoptera 123
Neotuberaphis (Pal) 54
Nicotiana tabacum 127
Nipponaphis (Pergande) 54, 194

O

Oedisiphum (Goot) 51
Okra 24
Onion 8
Ovatus (Goot) 51

P

Paddy 14
Paezoskia (Mordvilko) 51
Paniculum 13
Paramyzus Takahashi 117

Paraoregma (Ghosh) 54
Paratichosiphum alnicola Basu 152
Paratrichosiphum 153
Pea 8, 21
Peach 8, 19, 23, 27
Pennisetum 13
Pentalonia (Coquerel) 51
Pentalonia keli sp.nov. 111, 112, 115
Pentalonia nigronervosa 29
Pentalonia nigronervosa Coquerel 110, 114
Perillaphis (Takahashi) 51
Pisdium guajava 156, 160, 172
Pisum sativum Linn. 97
Pleotrichophorus (Borner) 51
Plum 27
Potato 8, 19, 22, 23, 27
Pseudaphis (Lambers) 51
Pseudoacythosiphon (Ghosh) 51
Pseudoastegopteryx (Ghosh) 54
Pseudoregma (Doncaster) 54
Psidium guavjava 3
Psyllids 2
Pterowea plagiophleps 199
Ptumomyzus 123
Pulses 8

Q

Quercus sp. 172

R

Radish 25
Rhodobium (Lambers) 51
Rhopalosiphonius (Baker) 51
Rhopalosiphum cashivi sp.nov. 80, 81, 88
Rhopalosiphum Koch 51, 79
Rhopalosiphum kolhapurensis sp.nov. 84, 85, 88
Rhopalosiphum maidis Fitch 83
Rhopalosiphum padi Linnaeus 87
Rhophalosiphum maidis Ashmed 2, 11
Rod 27
Rosa 22
Rose multiform Thumb 66, 144
Rubia 13

S

S. cumunis 63, 66, 77
Saccharum officinarum 184
Schizaphis Borner 51
Schoutendenia (Rubsaamen) 53
Sciamyzus 123
Semiaphis (Goot) 51
Shinjia (Takahashi) 51
Sinomegoura (Takahashi) 52
Siphonophora artocarpi Westwood 162
Siphonophora circumflex Buckton 145
Siphonophora Kach. 163
Solanum melongena 138
Sopolia Zeylan 201
Sorghum 8, 13
Sorghum vulgure Pers. 2, 13, 71
Sprout 25
Subovatomyus (Basu) 52
Subtakecallis brevisetosus 37
Sugarcane 8, 11, 13, 14

Sumatraphis (Takahashi) 53
Swede 25
Sweet potato 23
Syrphid 122, 185
Syrphid larvae 73
Syrphid fly 17

T

Taiwanomyzus (Tao) 52
Tamarind 31
Tea 31
Teberaphis (Takahashi) 54
Tecoma capensis 22
Tectona grandis 138
Terminalia chebula 63
Terminalia paniculata 110
Terminalia tomentosa 66, 131
Tetraneura javensis 14
Therioaphis sexmaculatus 21
Thoracaphis (Goot) 54
Tiger beetle *Eudema angulatum* 179
Tiger beetles 138
Tobacco 23
Tomato 8
Toxoptera ambi sp.nov 74, 75, 78
Toxoptera aurantii Boyer de Fonscolombe 31, 79
Toxoptera citricidus 30
Toxoptera Koch 51, 73
Triassoaphids cubitus 34
Tricaudatus (Narzikulvov) 52
Trichosiphonaphis (Takahashi) 52
Trichosiphum Pergande 163
Tubaphis 123
Tuberoaphis (Tao) 52
Tuberosephalus (Shinji) 52

U

Udid 21
Unisitobion 139

V

Vernia sp. 151, 162, 179
Vesiculaphis (Guercio) 52
Vigna catjang Walp. 3, 58, 63

W

Walgreniella (Lambers) 52
Wals 21
wheat 8, 12

X

Xenomyzus (Aizenberg) 52

Z

Z. zijuba 8
Zea mays 3
Zingiber 29